Zong Woo Geem (Ed.)

Harmony Search Algorithms for Structural Design Optimization

Studies in Computational Intelligence, Volume 239

Editor-in-Chief
Prof. Janusz Kacprzyk
Systems Research Institute
Polish Academy of Sciences
ul. Newelska 6
01-447 Warsaw
Poland
E-mail: kacprzyk@ibspan.waw.pl

Further volumes of this series can be found on our homepage: springer.com

Vol. 218. Maria do Carmo Nicoletti and Lakhmi C. Jain (Eds.)
Computational Intelligence Techniques for Bioprocess Modelling, Supervision and Control, 2009
ISBN 978-3-642-01887-9

Vol. 219. Maja Hadzic, Elizabeth Chang, Pornpit Wongthongtham, and Tharam Dillon
Ontology-Based Multi-Agent Systems, 2009
ISBN 978-3-642-01903-6

Vol. 220. Bettina Berendt, Dunja Mladenic, Marco de de Gemmis, Giovanni Semeraro, Myra Spiliopoulou, Gerd Stumme, Vojtech Svatek, and Filip Zelezny (Eds.)
Knowledge Discovery Enhanced with Semantic and Social Information, 2009
ISBN 978-3-642-01890-9

Vol. 221. Tassilo Pellegrini, Sören Auer, Klaus Tochtermann, and Sebastian Schaffert (Eds.)
Networked Knowledge - Networked Media, 2009
ISBN 978-3-642-02183-1

Vol. 222. Elisabeth Rakus-Andersson, Ronald R. Yager, Nikhil Ichalkaranje, and Lakhmi C. Jain (Eds.)
Recent Advances in Decision Making, 2009
ISBN 978-3-642-02186-2

Vol. 223. Zbigniew W. Ras and Agnieszka Dardzinska (Eds.)
Advances in Data Management, 2009
ISBN 978-3-642-02189-3

Vol. 224. Amandeep S. Sidhu and Tharam S. Dillon (Eds.)
Biomedical Data and Applications, 2009
ISBN 978-3-642-02192-3

Vol. 225. Danuta Zakrzewska, Ernestina Menasalvas, and Liliana Byczkowska-Lipinska (Eds.)
Methods and Supporting Technologies for Data Analysis, 2009
ISBN 978-3-642-02195-4

Vol. 226. Ernesto Damiani, Jechang Jeong, Robert J. Howlett, and Lakhmi C. Jain (Eds.)
New Directions in Intelligent Interactive Multimedia Systems and Services - 2, 2009
ISBN 978-3-642-02936-3

Vol. 227. Jeng-Shyang Pan, Hsiang-Cheh Huang, and Lakhmi C. Jain (Eds.)
Information Hiding and Applications, 2009
ISBN 978-3-642-02334-7

Vol. 228. Lidia Ogiela and Marek R. Ogiela
Cognitive Techniques in Visual Data Interpretation, 2009
ISBN 978-3-642-02692-8

Vol. 229. Giovanna Castellano, Lakhmi C. Jain, and Anna Maria Fanelli (Eds.)
Web Personalization in Intelligent Environments, 2009
ISBN 978-3-642-02793-2

Vol. 230. Uday K. Chakraborty (Ed.)
Computational Intelligence in Flow Shop and Job Shop Scheduling, 2009
ISBN 978-3-642-02835-9

Vol. 231. Mislav Grgic, Kresimir Delac, and Mohammed Ghanbari (Eds.)
Recent Advances in Multimedia Signal Processing and Communications, 2009
ISBN 978-3-642-02899-1

Vol. 232. Feng-Hsing Wang, Jeng-Shyang Pan, and Lakhmi C. Jain
Innovations in Digital Watermarking Techniques, 2009
ISBN 978-3-642-03186-1

Vol. 233. Takayuki Ito, Minjie Zhang, Valentin Robu, Shaheen Fatima, and Tokuro Matsuo (Eds.)
Advances in Agent-Based Complex Automated Negotiations, 2009
ISBN 978-3-642-03189-2

Vol. 234. Aruna Chakraborty and Amit Konar
Emotional Intelligence, 2009
ISBN 978-3-540-68606-4

Vol. 235. Reiner Onken and Axel Schulte
System-Ergonomic Design of Cognitive Automation, 2009
ISBN 978-3-642-03134-2

Vol. 236. Natalio Krasnogor, Belén Melián-Batista, José A. Moreno-Pérez, J. Marcos Moreno-Vega, and David Pelta (Eds.)
Nature Inspired Cooperative Strategies for Optimization (NICSO 2008), 2009
ISBN 978-3-642-03210-3

Vol. 237. George A. Papadopoulos and Costin Badica (Eds.)
Intelligent Distributed Computing III, 2009
ISBN 978-3-642-03213-4

Vol. 238. Li Niu, Jie Lu, and Guangquan Zhang
Cognition-Driven Decision Support for Business Intelligence, 2009
ISBN 978-3-642-03207-3

Vol. 239. Zong Woo Geem (Ed.)
Harmony Search Algorithms for Structural Design Optimization, 2009
ISBN 978-3-642-03449-7

Zong Woo Geem (Ed.)

Harmony Search Algorithms for Structural Design Optimization

Dr. Zong Woo Geem
1650 Research Blvd.
TA1105
Rockville, Maryland 20850
USA
E-mail: ZWGEEM@GMAIL.COM

ISBN 978-3-642-26052-0 e-ISBN 978-3-642-03450-3

DOI 10.1007/978-3-642-03450-3

Studies in Computational Intelligence ISSN 1860-949X

Softcover reprint of the hardcover 1st edition 2010

Typeset & Cover Design: Scientific Publishing Services Pvt. Ltd., Chennai, India.

Printed in acid-free paper

9 8 7 6 5 4 3 2 1

springer.com

Preface

Various structures, such as buildings, bridges, and paved roads play an important role in our lives. However, these construction projects require large expenditures. Designing infrastructure cost-efficiently while satisfying all necessary design constraints is one of the most important and difficult tasks for a structural engineer. Traditionally, mathematical gradient-based optimization techniques have been applied to these designs. However, these gradient-based methods are not suitable for discrete design variables such as factory-made cross sectional area of structural members.

Recently, researchers have turned their interest to phenomenon-mimicking optimization techniques because these techniques have proved able to efficiently handle discrete design variables. One of these techniques is harmony search, an algorithm developed from musical improvisation that has been applied to various structural design problems and has demonstrated cost-savings. This book gathers all the latest developments relating to the application of the harmony search algorithm in the structural design field in order for readers to efficiently understand the full spectrum of the algorithm's potential and to easily apply the algorithm to their own structural problems.

This book contains six chapters with the following subjects: standard harmony search algorithm and its applications by Lee; standard harmony search algorithm for steel frame design by Degertekin; adaptive harmony search algorithm and its applications by Saka and Hasançebi; harmony particle swarm algorithm and its applications by Li and Liu; hybrid algorithm of harmony search, particle swarm & ant colony for structural design by Kaveh and Talatahari; and parameter calibration of viscoelastic and damage functions by Mun and Geem.

As an editor of this book, I would like to express profound thanks to reviewers and proofreaders including Mike Dreis, John Galuardi, Teresa Giral, Sanghun Kim, and Ronald Wiles. Also, I would like to share the joy of the publication with my family (Joseph, Catarina, Victoria, Sophia, Michelle...).

Zong Woo Geem
Editor

Preface

Contents

Standard Harmony Search Algorithm for Structural Design Optimization

Kang Seok Lee[1]

Abstract. Most engineering optimization algorithms are based on numerical linear and nonlinear programming methods that require substantial gradient information and usually seek to improve the solution in the neighborhood of a starting point. These algorithms, however, reveal a limited approach to complicated real-world optimization problems. If there is more than one local optimum in the problem, the result may depend on the selection of an initial point, and the obtained optimal solution may not necessarily be the global optimum. The computational drawbacks of numerical methods have forced researchers to rely on meta-heuristic algorithms based on simulations to solve optimization problems. This chapter describes a basic harmony search (HS) meta-heuristic algorithm-based approach for optimizing the size and configuration of structural systems with both discrete and continuous design variables. This basic HS algorithm is conceptualized using the musical process of searching for a perfect state of harmony. It uses a stochastic random search instead of a gradient search so that derivative information is unnecessary. Various truss examples, including large-scale trusses under multiple loading conditions, are introduced to demonstrate the effectiveness and robustness of the basic harmony search algorithm-based methods, as compared to existing structural optimization techniques. The results indicate that the HS technique is a powerful search and optimization method for solving structural engineering problems compared to conventional mathematical methods or genetic algorithm-based approaches.

1 Introduction

During the last four decade, many mathematical programming methods, such as linear, nonlinear, and dynamic programming, have been developed and frequently used to solve optimization problems. These optimization methods provide a useful strategy to obtain global optima in simple and ideal models. However, many optimization problems, including those in structural engineering, are very complex in nature and quite difficult to solve using these methods. In linear programming, errors are inevitable when a linear relationship is used to model nonlinear real problems. In dynamic programming, an increase in the number of variables exponentially increases the number of required evaluations of the recursive functions. In nonlinear programming, the solution algorithm may not find the optimum if the functions used in computations are not differentiable; also, the selection of the initial starting values is important to ensure that the algorithm will converge to the global optimum and not a local optimum.

[1] School of Architecture, Chonnam National University, Gwangju, South Korea
Email: kslnist@chonnam.ac.kr

Z.W. Geem (Ed.): Harmony Search Algo. for Structural Design Optimization, SCI 239, pp. 1–49.
springerlink.com

Since the 1970s, many heuristic optimization algorithms that combine rules and randomness imitating natural phenomena have been devised to solve difficult optimization problems. These algorithms include simulated annealing, tabu search, and evolutionary algorithms. In 1983, Kirkpatrick *et al.* [1] proposed the innovative idea of a simulated annealing algorithm, which is based on an analogy to the physical annealing process. They modeled their approach after the stochastic thermal equilibrium process proposed by Metropolis *et al.* [2] to solve a classic combinatorial optimization problem (the traveling salesperson problem); good results were obtained. Tabu search is an iterative procedure for solving discrete combinatorial optimization problems that was originally suggested by Glover [3]. The basic idea of this algorithm is to explore the search space of all feasible solutions using a sequence of moves. A move from one solution to another results in the best available solution. However, to escape from local optima and to prevent cycling, some moves are classified as forbidden or tabu. Tabu moves are based on the history of the move sequence. Evolutionary algorithms, which are based on a principle of evolution (survival of the fittest) and imitate some natural phenomena (genetic inheritance), are basically composed of four heuristic algorithms: genetic algorithms, evolution strategies, evolutionary programming, and genetic programming.

Genetic algorithms are search algorithms based on natural selection and the mechanisms of population genetics. The theory was proposed by Holland [4] and further developed by Goldberg [5] and others. Simple genetic algorithms are comprised of three operators; reproduction, crossover, and mutation. The main characteristic of genetic algorithms is the simultaneous evaluation of many solutions, which differs from mathematical optimization or other heuristic methods such as simulated annealing or tabu searches. These and other similar methods evaluate only one solution at each iteration. This feature is an advantage, enabling a wide search and potentially avoiding convergence to a non-global optimum. Evolution strategies were developed to solve parameter optimization problems [6], in which a deterministic ranking is used to select a basic set of solutions for a new trial [7]. Evolutionary programming, which was originally developed by Fogel *et al.* [8], described the evolution of finite state machines to solve prediction tasks. The state transition tables in these machines are modified by uniform random mutations on the corresponding alphabet. The algorithms utilize selection and mutation as the main operators, and the selection process is a stochastic tournament. Genetic programming, which is an extension of genetic algorithms, was developed relatively recently by Koza [9]. He suggested that the desired program should itself evolve during the evolution process.

The simulation-based heuristic methods described above have powerful searching abilities that can occasionally overcome several deficiencies of the mathematical methods. There are numerous applications of these heuristic optimization methods to various engineering optimization problems. Especially in the last decade, genetic algorithms have been used to solve various structural optimization problems and good results have been obtained. These include researches by Adeli and Cheng [10], Hajela [11], Jenkins [12-15], Grierson and Pak [16], Oshaki [17], Rajan [18], Yang and Soh [19], Galante [20], Rajeev and Krishnamoorthy [21, 22], Koumousis and Georgious [23], Hajela and Lee [24], Adeli and Kumar [25],

Wu and Chow [26, 27], Soh and Yang [28], Camp *et al.* [29], Shrestha and Ghaboussi [30], Erbatur *et al.* [31], and Sarma and Adeli [32]. Compared to other optimization methods, genetic algorithms have the advantage of imposing fewer mathematical requirements for solving the problems and being very effective at performing global searches. However, the characteristics that make genetic algorithms robust also make them computationally intensive, requiring high computing costs; hence they are slower than other methods (Hajela [11], Haftka *et al.* [33], Jenkins [15], and Soh and Yang [28]). From the point of view of structural designers, long computing times are not acceptable. However, there are still some possibilities of devising new heuristic algorithms based on analogies with natural or artificial phenomena.

Geem *et al.* [34] developed a basic harmony search (BHS) heuristic optimization algorithm that was based on an analogy with the process of music improvisation. The harmony in music is analogous to the optimization solution vector and the musician's improvisations are analogous to local and global search schemes in optimization techniques. Although the BHS algorithm is a comparatively simple method, it has been successfully applied to various optimization problems including the traveling salesperson problem, the layout of pipe networks, pipe capacity design in water supply networks, hydrologic model parameter calibrations, and optimal school bus routings.

This chapter describes a BHS algorithm-based approach for optimizing the size and configuration of structural systems with both discrete and continuous design variables. Various truss examples, including large-scale trusses under multiple loading conditions, are also presented to demonstrate the effectiveness and robustness of the BHS algorithm-based methods, as compared to existing structural optimization techniques. Although the proposed approach is applied to truss structures, it is a general optimization procedure that can be easily used for other types of structures, such as frame structures, plates, and shells.

2 Statement of the Optimization Design Problem

Design objectives that can be used to measure design quality include minimum construction cost, minimum life cycle cost, minimum weight, and maximum stiffness, as well as many others. Typically, the design is limited by constraints such as the choice of material, feasible strength, displacements, eigen-frequencies, load cases, support conditions, and technical constraints (*e.g.*, type and size of available structural members and cross sections, *etc*). Hence, one must decide which parameters can be modified during the optimization process; these parameters then become the optimization variables. Usually, structural optimization problems involve searching for the minimum of the structural weight. This minimum weight design is subjected to various constraints with respect to performance measures, such as stresses and displacements, and also restricted by practical minimum cross-sectional areas or dimensions of the structural members or components. If the design variables can be varied continuously in the optimization, the problem is termed “continuous”; while if the design variables represent a selection from a set of parts, the problem is considered “discrete”.

On the other hand, isotropic structures can be usually described by three different types of design variables: (1) sizing variables, (2) geometric variables, and (3) topological variables. Sizing optimization is concerned with determining the cross section size. Configuration optimization searches for a set of geometric and sizing variables using a given topology. A selection from various structural types can be made in topology optimization. In general, size and geometric variables are frequently used to solve structural optimization design problems.

This chapter considers a BHS algorithm-based approach for optimizing the size and configuration of structural systems with both discrete and continuous design variables.

2.1 Continuous Size and Configuration Optimization

The continuous size optimization of structural systems involves arriving at optimum values for member continuous cross-sectional areas $\boldsymbol{A}$ that minimize an objective function $f(\boldsymbol{x})$, *i.e.*, the structural weight W. The continuous configuration optimization involves simultaneously arriving at optimum values for the nodal coordinates $\boldsymbol{R}$ and member cross-sectional areas $\boldsymbol{A}$ that minimize an objective function. For a given topology, the configuration optimization problem is generally considered to be more difficult, but it is also a more important task than pure size optimization because of the potential for much larger savings. Both minimum designs must satisfy q inequality constraint functions that limit the design variable sizes and the structural responses. Thus, the problems can be stated mathematically, as minimizing the structural weight:

$$\text{Minimize } f(\boldsymbol{x}) = W(\boldsymbol{A}) \text{ or } W(\boldsymbol{R}, \boldsymbol{A}) = \gamma \sum_{i=1}^{n} L_i A_i \quad (1)$$

$$\text{subjected to } G_j^l \leq G_j(\boldsymbol{A}) \text{ or } G_j(\boldsymbol{R}, \boldsymbol{A}) \leq G_j^u, j = 1,2,\ldots,q \quad (2)$$

where $f(\boldsymbol{x})$ is an objective function, $\boldsymbol{x}$ is the continuous set of each design variable, L_i = the member length and γ = the material density.

For the continuous size optimization method presented in this chapter, the upper and lower bounds on the constraint function $G_j(\boldsymbol{A})$ or $G_j(\boldsymbol{R}, \boldsymbol{A})$ in Eq. (2) include the following: (a) nodal coordinates ($R_j^l \leq R_i \leq R_j^u$, $i = 1,\ldots, m$); (b) member continues cross sections ($A_j^l \leq A_i \leq A_j^u$, $i = 1,\ldots, n$); (c) member stresses ($\sigma_i^l \leq \sigma_i \leq \sigma_i^u$, $i = 1,\ldots,n$); (d) nodal displacements ($\delta_i^l \leq \delta_i \leq \delta_i^u$, $i = 1,\ldots,m$); and (e) member buckling stress ($\sigma_i^{cr} \leq \sigma_i \leq 0$, $i = 1,\ldots,n$). Here, σ_i and δ_i are the member stresses and nodal displacements, respectively, calculated from the structural analysis; R_i^l, R_i^u, σ_i^l, σ_i^u, δ_i^l, δ_i^u, and σ_i^{cr} are the constraint limitations prescribed for optimization design purposes. The nodal coordinate constraints are required only for the continuous configuration optimization.

2.2 Discrete Size and Discrete-Continuous Configuration Optimization

The discrete size optimization of structural systems involves arriving at optimum values for discrete member cross-sectional areas $\boldsymbol{A}$ that minimize an objective function $f(\boldsymbol{x})$, *i.e.*, the structural weight W. Discrete-continuous configuration optimization involves simultaneously arriving at optimum values for continuous nodal coordinates $\boldsymbol{R}$ and discrete cross sections $\boldsymbol{A}$ that minimize the structural weight. Both minimum designs must satisfy q inequality constraint functions that limit the design variable sizes and the structural responses. The design problem is also expressed as Eqs. (1) and (2).

However, $\boldsymbol{x}$ is the discrete set of each design variable, $\boldsymbol{A} = (A_1, A_2,\ldots, A_n)^T$ is the sizing variable vector that consists of the cross-sectional areas chosen from a list of available discrete values, and $\boldsymbol{R} = (R_1, R_2,\ldots, R_m)^T$ is the continuous nodal coordinate variable vector. Also, $W(\boldsymbol{A})$ and $W(\boldsymbol{R}, \boldsymbol{A})$ are the objective functions (*i.e.*, the structural weight) for the discrete size or the discrete-continuous configuration optimizations, respectively, γ is the material density of each member, and A_i and L_i are the cross-sectional area and length of the ith member. $G_j(\boldsymbol{A})$ or $G_j(\boldsymbol{R}, \boldsymbol{A})$, shown in Eq. (2), are the inequality constraints for the discrete size or the discrete-continuous configuration optimizations, and G_j^l and G_j^u are the lower and the upper bounds on the constraints.

For the discrete and the discrete-continues configuration optimization methods presented in this chapter, the lower and upper bounds on the constraint function Eq. (2) include the following: (a) nodal coordinates $(R_i^l \le R_i \le R_i^u,\ i=1,\ldots,m)$; (b) member cross sections $(A_i(k),\ i=1,\ldots,n)$; (c) member stresses $(\sigma_i^l \le \sigma_i \le \sigma_i^u,\ i=1,\ldots,n)$; (d) nodal displacements $(\delta_i^l \le \delta_i \le \delta_i^u,\ i=1,\ldots,m)$; and (e) member buckling stresses $(\sigma_i^{cr} \le \sigma_i \le 0,\ i=1,\ldots,n)$. Here, σ_i and δ_i are the member stresses and nodal displacements, respectively, calculated from the structural analysis; R_i^l, R_i^u, σ_i^l, σ_i^u, δ_i^l, δ_i^u, and σ_i^{cr} are the constraint limitations prescribed for optimization design purposes; and $A_i(k)$ are the available discrete cross-sectional areas, *i.e.*, $A_i(1), A_i(2),\ldots, A_i(k)$ $(A_i(1) < A_i(2) < \ldots < A_i(k))$. The nodal coordinate constraints are required only for the discrete-continuous configuration optimization.

3 Basic Harmony Search Algorithm-Based Structural Optimization and Design Procedures

The penalty approach has frequently been employed to determine the fitness measure for the constrained optimization problems, described by Eqs. (1) and (2), because the optimum solution typically occurs at the boundary between the feasible and infeasible regions (Rajeev and Krishnamoorthy [21]; Wu and Chow [27]; Camp *et al.* [29]; Pezeshk *et al.* [35]; and Erbatur *et al.* [31]). However, to demonstrate the pure performance of the BHS algorithm-based methods proposed in this chapter, a rejecting strategy for the fitness measure was adopted, *i.e.*, the optimum solution was approached only from the feasible region.

Figure 1 showed the design procedure that was used to apply the BHS algorithm to the continuous size optimization, the continuous configuration optimizations, the discrete size optimization, and the discrete-continuous configuration optimization problems. The procedure can be divided into four steps, as follows.

3.1 Step 1: Initialization

The optimization problem is first specified as $W(\boldsymbol{A})$ or $W(\boldsymbol{R}, \boldsymbol{A})$ in Eq. (1). For continuous size optimization problems, *i.e.*, $W(\boldsymbol{A})$, the possible value bounds of the continuous design variables (A_i) , *i.e.*, $A_j^l \le A_i \le A_j^u$ are then initialized. For continuous configuration optimization problems, *i.e.*, $W(\boldsymbol{R}, \boldsymbol{A})$, the number of continuous geometric variables (R_i) and the possible value bounds of the continuous variables, *i.e.*, $R_i^l \le R_i \le R_i^u$ are initialized.

On the other hand, for discrete size optimization problems, *i.e.*, $W(\boldsymbol{A})$, the number of discrete design variables (A_i) and the set of available discrete values (D), *i.e.*, $D \in \{A_i(1), A_i(2),\ldots, A_i(k)\}$ $(A_i(1) < A_i(2) < \ldots < A_i(k))$ are then initialized. For discrete-continuous configuration optimization problems, *i.e.*, $W(\boldsymbol{R}, \boldsymbol{A})$, the number of continuous geometric variables (R_i) and the possible value bounds of the continuous variables, *i.e.*, $R_i^l \le R_i \le R_i^u$ are initialized, as well as the discrete design variables.

The BHS algorithm parameters that are required to solve the optimization problem are also specified in this step. These include the harmony memory size (number of solution vectors in the harmony search, HMS), harmony memory considering rate (HMCR), pitch adjusting rate (PAR), and termination criterion (maximum number of searches). The HMCR and the PAR are parameters that are used to improve the solution vector. Both are defined in Step 2. Subsequently, the "harmony memory" (HM) matrix, shown in Eq. (3), is randomly generated from the available continuous and discrete value set and/or the possible nodal coordinate bounds for the optimization problems. These sets are equal to the size of the HM (*i.e.*, HMS). Here, an initial HM is generated based on the FEM structural analysis results, subject to the constraint functions (Eq. [2]), and sorted by the objective function values (Eq. [1]).

$$\boldsymbol{HM} = \begin{bmatrix} x_1^1 & x_2^1 & \cdots & x_p^1 \\ x_1^2 & x_2^2 & \cdots & x_p^2 \\ \vdots & \vdots & \cdots & \vdots \\ x_1^{HMS} & x_2^{HMS} & \cdots & x_p^{HMS} \end{bmatrix} \begin{matrix} \Rightarrow f(\boldsymbol{x}^1) \\ \Rightarrow f(\boldsymbol{x}^2) \\ \Rightarrow \quad \vdots \\ \Rightarrow f(\boldsymbol{x}^{HMS}) \end{matrix} \tag{3}$$

In Eq. (3), $\boldsymbol{x}^1, \boldsymbol{x}^2,\ldots, \boldsymbol{x}^{HMS}$ and $f(\boldsymbol{x}^1), f(\boldsymbol{x}^2),\ldots, f(\boldsymbol{x}^{\mathrm{HMS}})$ show each solution vector for design variables ($\boldsymbol{A}$ or $\boldsymbol{R}$ and $\boldsymbol{A}$) and the corresponding objective function value (the structural weight), respectively.

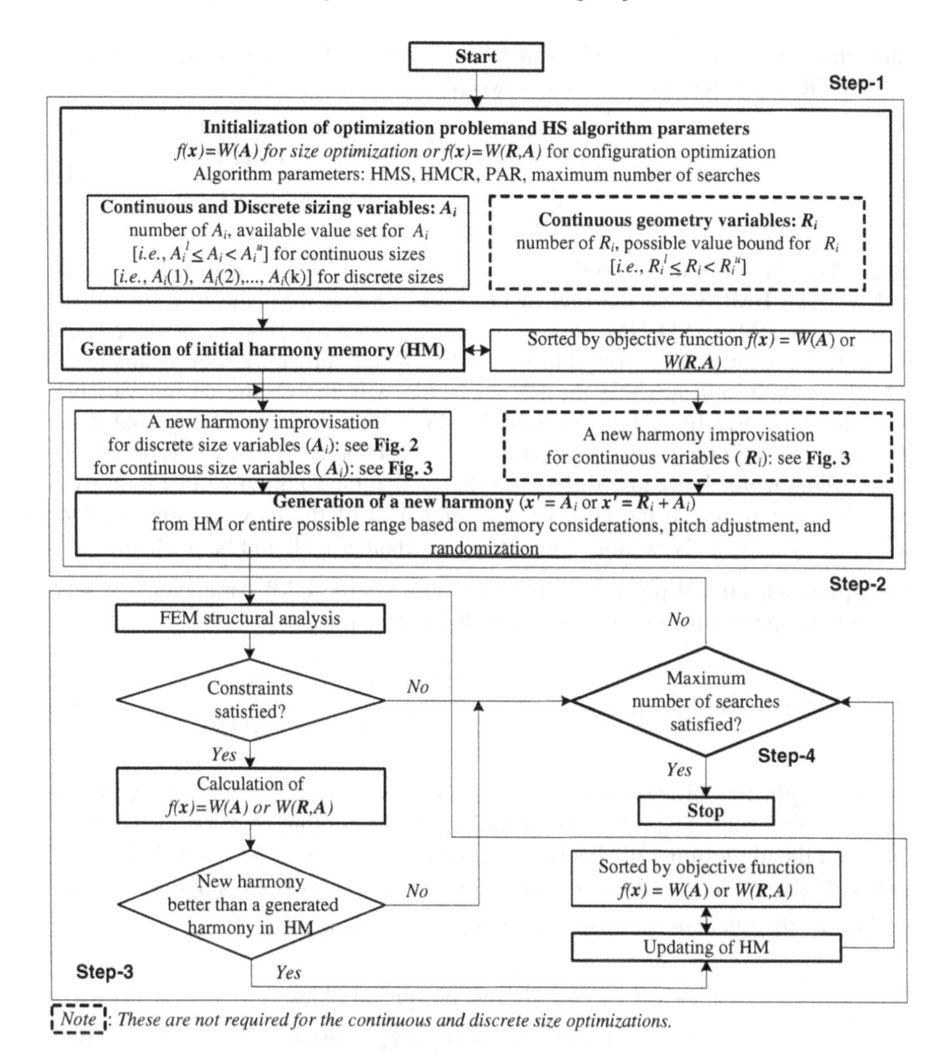

Fig. 1 BHS algorithm-based Structural Optimization Design Procedure

3.2 *Step 2: Generation of a New Harmony*

In the HS algorithm, a new harmony vector, $\boldsymbol{x}' = (x_1', x_2', ..., x_p')$, is improvised from either the initially generated HM or the entire possible range of values. The new harmony improvisation proceeds based on memory considerations, pitch adjustments, and randomization.

In the memory consideration process, the value of the first design variable (x_1') for the new vector is chosen from any value in the specified HM range $\{x_1^1, x_1^2, \cdots, x_1^{HMS}\}$. Values of the other decision variables (x_i') are chosen in the same manner. Here, the possibility that a new value will be chosen is indicated by the HMCR parameter, which varies between 0 and 1 as follows:

$$x_i' \leftarrow \begin{cases} x_i' \in \{x_i^1, x_i^2, \ldots, x_i^{HMS}\} & w.p. \quad HMCR \\ x_i' \in \boldsymbol{X}_i & w.p. \quad (1\text{-}HMCR) \end{cases} \tag{4}$$

where $\boldsymbol{X}_i$ is the set of the possible range of values for each design variable (***A*** or ***R*** and ***A***). The HMCR sets the rate of choosing a value from the historic values stored in the HM, and (1-HMCR) sets the rate of randomly choosing a value from the entire possible range of values (randomization process). For example, a HMCR of 0.90 indicates that the HS algorithm will choose the design variable value from historically stored values in the HM with a 90% probability, and from the entire possible range of values with a 10% probability. A HMCR value of 1.0 is not recommended, because there is no chance that the solution will be improved by values not stored in the HM. Every component of the new harmony vector, $\boldsymbol{x}' = (x_1', x_2', \ldots, x_p')$, is examined to determine whether it should be pitch-adjusted using pitch adjustment process. This procedure uses the PAR parameter that sets the rate of adjusting the pitch chosen from the HM as follows:

$$\text{Pitch adjusting decision for } x_i' \leftarrow \begin{cases} Yes & w.p. \quad PAR \\ No & w.p. \quad (1-PAR) \end{cases} \tag{5}$$

The pitch adjusting process is performed only after a value has been chosen from the HM. The value (1-PAR) sets the rate of doing nothing. A PAR of 0.3 indicates that the algorithm will choose a neighboring value with 30% × HMCR probability. If the pitch adjustment decision for x_i' is Yes, and x_i' is assumed to be $x_i(l)$, *i.e.*, the *l*-th element in $\boldsymbol{X}_i$, the pitch-adjusted value of $x_i(l)$ is

$$\begin{aligned} &x_i' \leftarrow x_i(l + c) \text{ for discrete design variables} \\ &x_i' \leftarrow x_i' + \alpha \text{ for continuous design variables} \end{aligned} \tag{6}$$

where c is the neighboring index, $c \in \{-1, 1\}$; α is the value of $bw \times u(-1, 1)$; bw is an arbitrary distance bandwidth for the continuous variable; and $u(-1, 1)$ is a uniform distribution between -1 and 1. Detailed flowcharts for the new harmony discrete and continuous search strategies based on the BHS heuristic algorithm are given in Figs. 2 and 3, respectively. Note that the HMCR and PAR parameters introduced in the harmony search help the algorithm find globally and locally improved solutions.

3.3 Step 3: Fitness Measure and HM Update

The new harmony improvised in Step 2 is analyzed using a FEM structural analysis method, and its fitness is determined using a rejection strategy based on the

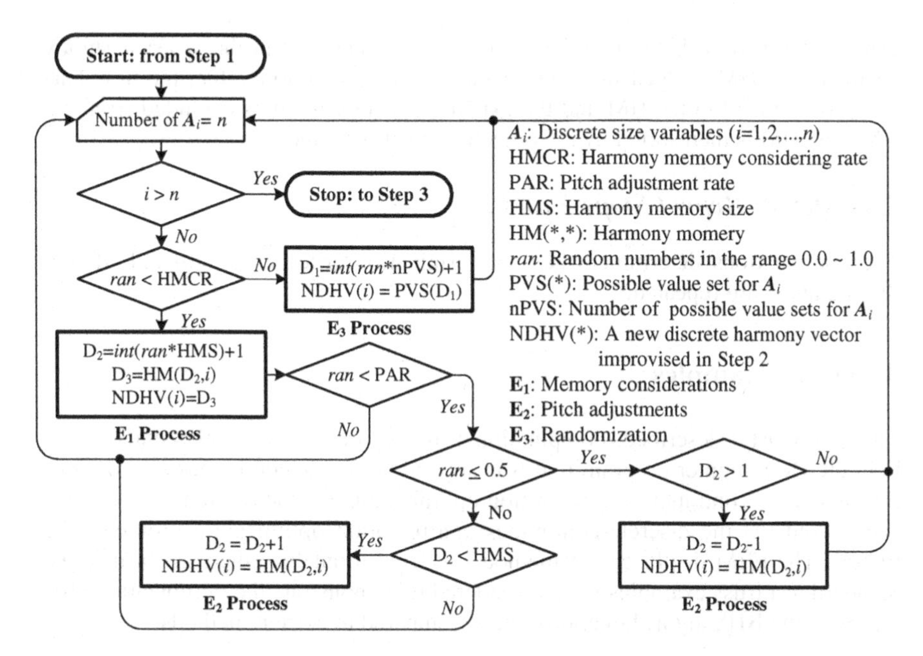

Fig. 2 A New Harmony Improvisation Flowchart For Discrete Variables (Step 2)

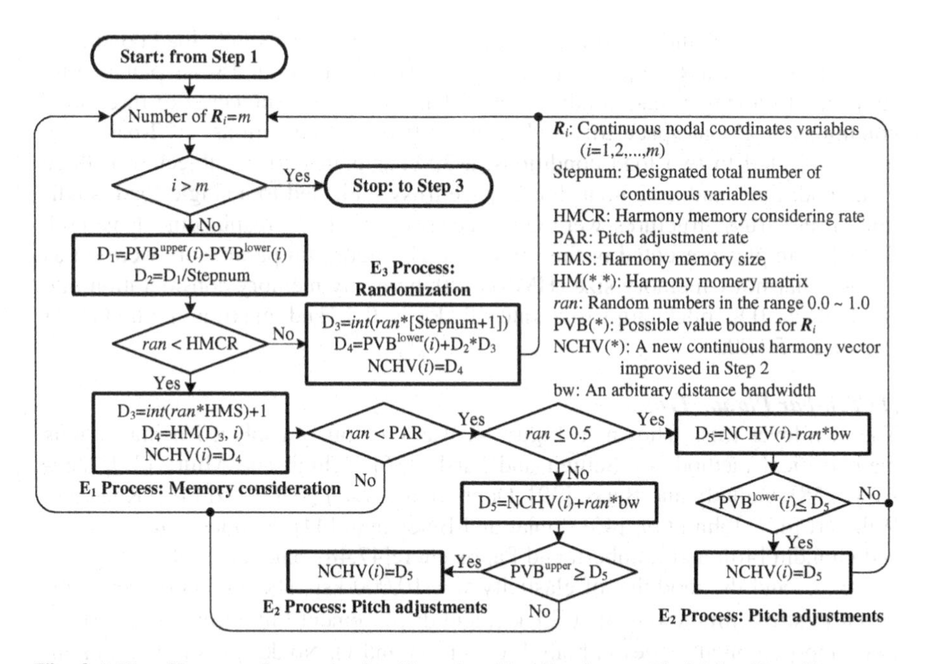

Fig. 3 A New Harmony Improvisation Flowchart For Continuous Variables (Step 2)

constraint function. If the new harmony vector is better than the worst harmony vector in the HM, judged in terms of the objective function value, the new harmony is included in the HM and the existing worst harmony is excluded from the HM. The HM is then sorted by the objective function value.

3.4 Step 4: Repeat Steps 2 and 3

The computations determine when the termination criterion is satisfied. If not, Steps 2 and 3 are repeated.

4 Truss Examples

The previously described computational procedures were implemented in a FORTRAN computer program that was applied to (1) the continuous size optimization, (2) the continuous configuration optimization, (3) the discrete size optimization, and (4) the discrete-continuous configuration optimization problems for trusses. The FEM displacement method was used to analyze the truss structures. Standard test truss examples were considered to demonstrate the optimization efficiency of the BHS algorithm approach, as compared to current methods.

4.1 Continuous Size Optimization Examples

These examples include a 10-bar planar truss subjected to a single load condition, a 17-bar planar truss subjected to a single load condition, an 18-bar planar truss subjected to a single load condition, a 22-bar space truss subjected to three load conditions, a 25-bar space truss subjected to two load conditions, a 72-bar space truss subjected to two load conditions, a 200-bar planar truss subjected to three load conditions, and a 120-bar dome space truss subjected to a single load condition. These truss structures were analyzed using the FEM displacement method. For all examples presented in this study, the HS algorithm parameters were set as follows: harmony memory size (HMS) = 20, harmony memory consideration rate (HMCR) = 0.8, pitch adjusting rate (PAR) = 0.3, and maximum number of searches = 50,000.

(1) Ten-bar Planar Truss

The cantilever truss, shown in Figure 4, was previously analyzed using various mathematical methods by Schmit and Farshi [36], Schmit and Miura [37], Venkayya [38], Gellatly and Berke [39], Dobbs and Nelson [40], Rizzi [41], Khan and Willmert [42], John *et al.* [43], Sunar and Belegundu [44], Stander *et al.* [45], Xu and Grandhi [46], and Lamberti and Pappalettere[47,48]. The material density was 0.1 lb/in.^{3} and the modulus of elasticity was 10,000 ksi. The members were subjected to stress limitations of ± 25 ksi, and displacement limitations of ± 2.0 in. were imposed on all nodes in both directions (x and y). No design-variable linking was used; thus there are ten independent design variables. In this example, two cases were considered: Case 1, in which the single loading condition of $P_1 = 100$ kips and $P_2 = 0$ was considered; and Case 2, in which the single loading condition of P_1 = 150 kips and P_2 = 50 kips was considered. The minimum cross-sectional area of the members was 0.1 in.^{2}.

The HS algorithm-based method was applied to each case. It found 20 different solution vectors (*i.e.*, the values of the ten design-independent variables) after 50,000 searches for both cases. Tables 1 and 2 give the best solution vector for Cases 1 and 2, respectively, and also provide a comparison between the optimal design results reported in the literature and the present work. For Case 1, the best HS solution vector was (30.15, 0.102, 22.71, 15.27, 0.102, 0.544, 7.541, 21.56, 21.45, 0.100) and the corresponding objective function value (minimum weight of the structure) was 5,057.88 lb. For Case 2, the best HS solution vector was (23.25, 0.102, 25.73, 14.51, 0.100, 1.977, 12.21, 12.61, 20.36, 0.100) and the corresponding objective function value was 4,668.81 lb. The best solutions for Cases 1 and 2 were obtained after approximately 20,000 and 15,000 searches, respectively. These searches took three and two minutes on a Pentium 600 MHz computer. The HS algorithm results for each case were better optimized than those by previous mathematical studies reported in the literature.

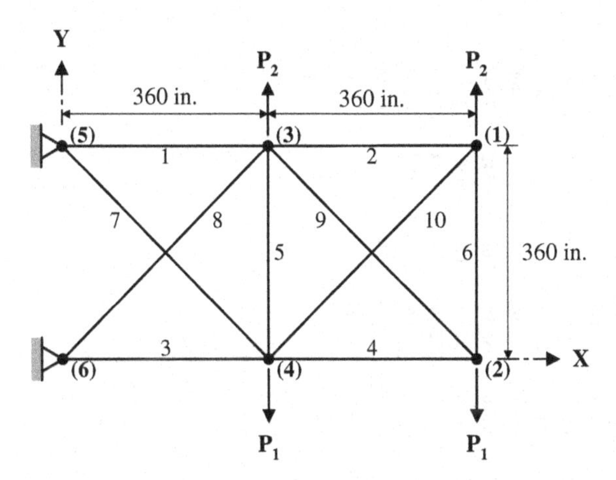

Fig. 4 Ten-bar planar truss

(2) Seventeen-bar Planar Truss

The 17-bar planar truss, shown in Figure 5, has been studied by Khot and Berke [49] and Adeli and Kumar [25]. The material density was 0.268 lb/in.^{3} and the modulus of elasticity was 30,000 ksi. The members were subjected to stress limitations of ±50 ksi, and displacement limitations of ±2.0 in. were imposed on all nodes in both directions (*x* and *y*). The single vertical downward load of 100 kips at node 9 was considered. No design-variable linking was used; thus there are seventeen independent design variables. The minimum cross-sectional area of the members was 0.1 in.^{2}. The HS algorithm was applied to the 17-bar planar truss with seventeen independent design variables.

Table 1 Optimal design comparison for the 10-bar planar truss (Case 1)

Variables		Optimal cross-sectional areas (in.^{2})													
		Schmit And Farshi [36]	Schmit and Miura [37] NEW-SUMT	Schmit and Miura [37] CON-MIN	Venkayya [38]	Gellatly and Berke [39]	Dobbs and Nelson [40]	Rizzi [41]	Khan and Willmert [42]	Sunar and Belegu-ndu [44]	Stand-er *et al.* [45]	Xu and Grandhi [46]	Lamberti and Pappalettere LEAML [47]	Lamberti and Pappalettere LESLP [48]	***This work***
1	A_1	33.43	30.67	30.57	30.42	31.35	30.50	30.73	30.98						***30.15***
2	A_2	0.100	0.100	0.369	0.128	0.100	0.100	0.100	0.100						***0.102***
3	A_3	24.26	23.76	23.97	23.41	20.03	23.29	23.93	24.17						***22.71***
4	A_4	14.26	14.59	14.73	14.91	15.60	15.43	14.73	14.81						***15.27***
5	A_5	0.100	0.100	0.100	0.101	0.140	0.100	0.100	0.100	*	*	*	*	*	***0.102***
6	A_6	0.100	0.100	0.364	0.101	0.240	0.210	0.100	0.406						***0.544***
7	A_7	8.388	8.578	8.547	8.696	8.350	7.649	8.542	7.547						***7.541***
8	A_8	20.74	21.07	21.11	21.08	22.21	20.98	20.95	21.05						***21.56***
9	A_9	19.69	20.96	20.77	21.08	22.06	21.82	21.84	20.94						***21.45***
10	A_{10}	0.100	0.100	0.320	0.186	0.100	0.100	0.100	0.100						***0.100***
Weight(lb)		5089.0	5076.85	5107.3	5084.9	5112.0	5080.0	5076.66	5066.98	5060.9	5060.85	5065.25	5060.96	5060.88	***5057.88***

Note: 1 in.^{2} = 6.452 cm^2, 1 lb = 4.45 N. * Unavailable

Table 2 Optimal design comparison for the 10-bar planar truss (Case 2)

Variables		Optimal cross-sectional areas (in.^{2})								
		Schmit and Farshi [36]	Schmit and Miura [37]		Venkayya [38]	Dobbs and Nelson [40]	Rizzi [41]	Khan and Willmert [42]	John *et al.* [43]	***This work***
			NEW-SUMT	CON-MIN						
1	A_1	24.29	23.55	23.55	25.19	25.81	23.53	24.72	23.59	***23.25***
2	A_2	0.100	0.100	0.176	0.363	0.100	0.100	0.100	0.10	***0.102***
3	A_3	23.35	25.29	25.20	25.42	27.23	25.29	26.54	25.25	***25.73***
4	A_4	13.66	14.36	14.39	14.33	16.65	14.37	13.22	14.37	***14.51***
5	A_5	0.100	0.100	0.100	0.417	0.100	0.100	0.108	0.10	***0.100***
6	A_6	1.969	1.970	1.967	3.144	2.024	1.970	4.835	1.97	***1.977***
7	A_7	12.67	12.39	12.40	12.08	12.78	12.39	12.66	12.39	***12.21***
8	A_8	12.54	12.81	12.86	14.61	14.22	12.83	13.78	12.80	***12.61***
9	A_9	21.97	20.34	20.41	20.26	22.14	20.33	18.44	20.37	***20.36***
10	A_{10}	0.100	0.100	0.100	0.513	0.100	0.100	0.100	0.10	***0.100***
Weight (lb)		4691.84	4676.96	4684.11	4895.60	5059.7	4676.92	4792.52	4676.93	***4668.81***

Note: 1 in.^{2} = 6.452 cm^2, 1 lb = 4.45 N. * Unavailable

The optimal results were compared to the earlier solutions reported by Khot and Berke [49] and Adeli and Kumar [25] in Table 3. Khot and Berke solved the problem using the optimality criterion method, and obtained a minimum weight of 2,581.89 lb. However, Adeli and Kumar solved the problem using a variant of genetic algorithms, obtaining a minimum weight of 2,594.42 lb. The HS heuristic algorithm-based technique found an optimum weight of 2,580.81 lb after approximately 20,000 searches that took less than three minutes. The optimal design

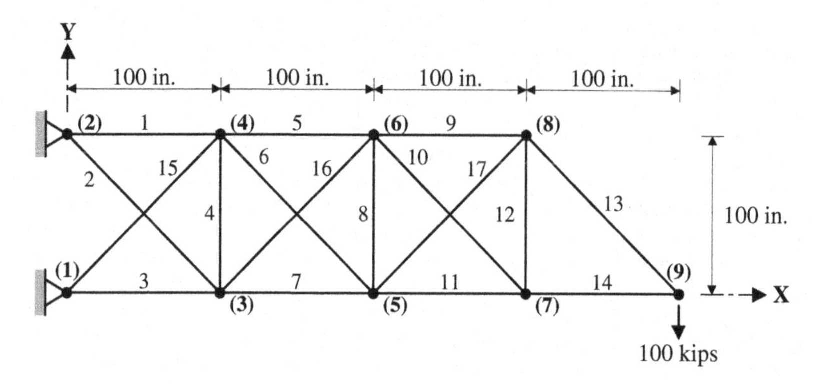

Fig. 5 Seventeen-bar planar truss

Table 3 Optimal design comparison for the 17-bar planar truss

Variables		Optimal cross-sectional areas (in.^{2})		
		Khot and Berke [49]	Adeli and Kumar [25]	***This work***
1	A_1	15.930	16.029	***15.821***
2	A_2	0.100	0.107	***0.108***
3	A_3	12.070	12.183	***11.996***
4	A_4	0.100	0.110	***0.100***
5	A_5	8.067	8.417	***8.150***
6	A_6	5.562	5.715	***5.507***
7	A_7	11.933	11.331	***11.829***
8	A_8	0.100	0.105	***0.100***
9	A_9	7.945	7.301	***7.934***
10	A_{10}	0.100	0.115	***0.100***
11	A_{11}	4.055	4.046	***4.093***
12	A_{12}	0.100	0.101	***0.100***
13	A_{13}	5.657	5.611	***5.660***
14	A_{14}	4.000	4.046	***4.061***
15	A_{15}	5.558	5.152	***5.656***
16	A_{16}	0.100	0.107	***0.100***
17	A_{17}	5.579	5.286	***5.582***
Weight lb)		2581.89	2594.42	***2580.81***

Note: 1 in.^{2} = 6.452 cm^2, 1 lb = 4.45 N.

obtained using the HS algorithm was slightly better than both of the previous design results.

(3) *Eighteen-bar Planar Truss*

The 18-bar cantilever planar truss, shown in Figure 6, was analyzed by Imai and Schmit [50] to obtain the optimal size variables. The material density was 0.1 lb/in.^{3} and the modulus of elasticity was 10,000 ksi. The members were subjected to stress limitations of ± 20 ksi. Also, an Euler bucking compressive stress limitation was imposed for truss member *i*, according to

$$_{b}\sigma_i = \frac{-KEA_i}{L_i^2} \tag{7}$$

where K = a constant determined from the cross-sectional geometry; E = the modulus of elasticity; and L_i = the member length. In this study, the buckling constant was taken to be K = 4. The single loading condition was a set of vertical loads with P = 20 kips acting on the upper nodal points of the truss, as illustrated in Figure 6. The cross-sectional areas of the members were linked into four groups, as follows: (1) $A_1=A_4=A_8=A_{12}=A_{16}$, (2) $A_2=A_6=A_{10}=A_{14}=A_{18}$, (3) $A_3=A_7=A_{11}=A_{15}$, and (4) $A_5=A_9=A_{13}=A_{17}$. The minimum cross sectional area was 0.1 in.^{2}. The HS algorithm-based method was applied to the 18-bar truss with four independent design variables.

The optimal results are compared to the earlier solutions reported by Imai and Schmit [50] in Table 4. Imai and Schmit solved the problem using the multiplier method, and obtained a minimum weight of 6,430.0 lb. The HS algorithm-based method found an optimum weight of 6,421.88 lb after approximately 2,000 searches that took less than one minute. The optimal design obtained using the HS algorithm was slightly better than the previous design obtained by Imai and Schmit.

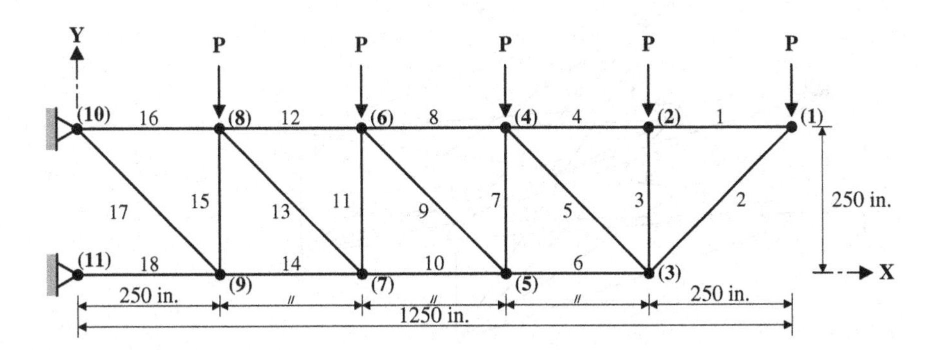

Fig. 6 Eighteen-bar planar truss.

Table 4 Optimal design comparison for the 18-bar planar truss

	Variables	Optimal cross-sectional areas (in.^{2}) Imai and Schmit [50]	*This work*
1	$A_1=A_4=A_8=A_{12}=A_{16}$	9.998	***9.980***
2	$A_2=A_6=A_{10}=A_{14}=A_{18}$	21.65	***21.63***
3	$A_3=A_7=A_{11}=A_{15}$	12.50	***12.49***
4	$A_5=A_9=A_{13}=A_{17}$	7.072	***7.057***
	Weight (lb)	6430.0	***6421.88***

Note: 1 in.^{2} = 6.452 cm^2, 1 lb = 4.45 N.

(4) Twenty-two-bar Space Truss

In the structure shown in Figure 7, each node is connected to every other node by a member, except for members between the fixed support nodes 5, 6, 7, and 8. The structure was previously studied by Khan and Willmert [42] and Sheu and Schmit [51] to determine the global optimum of trusses with vanishing members. In the example considered in this study, however, only the case with all groups of members (non-vanishing) was considered. The modulus of elasticity and the material density of all members were 10,000 ksi and 0.1 lb/in.^{3}, respectively. The twenty-two members were linked into seven groups, as follows: (1) $A_1 \sim A_4$, (2) $A_5 \sim A_6$, (3) $A_7 \sim A_8$, (4) $A_9 \sim A_{10}$, (5) $A_{11} \sim A_{14}$, (6) $A_{15} \sim A_{18}$, and (7) $A_{19} \sim A_{22}$. The truss members were subjected to the stress limitations shown in Table 5. Also, displacement constraints of ± 2.0 in. were imposed on all nodes in all directions. Three loading conditions described in Table 6 were considered, and a minimum member cross sectional area of 0.1 in.^{2} was enforced.

Table 7 lists the optimal values of the seven size variables obtained by the HS algorithm-based method, and compares them with earlier results reported by Khan and Willmert [42] and Sheu and Schmit [51]. The HS algorithm-based method

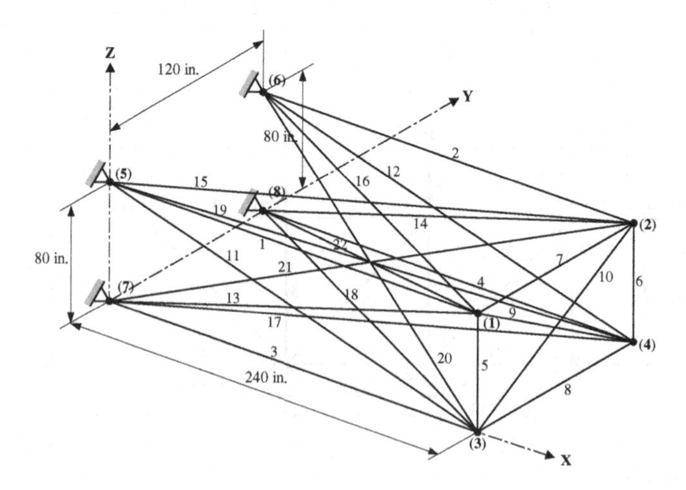

Fig. 7 Twenty-two-bar space truss.

achieves a design with a best solution vector of (2.588, 1.083, 0.363, 0.422, 2.827, 2.055, 2.044) and a minimum weight of 1,022.23 lb after approximately 10,000 searches. The optimal design obtained using the HS algorithm was slightly better than the results obtained by Sheu and Schmit [51], and had a minimum weight that was 1.2 % less than that obtained by Khan and Willmert [42].

Table 5 Member stress limitations for the 22-bar space truss

Variables		Compressive stress limitations (ksi)	Tensile stress limitations (ksi)
1	$A_1 \sim A_4$	24.0	36.0
2	$A_5 \sim A_6$	30.0	36.0
3	$A_7 \sim A_8$	28.0	36.0
4	$A_9 \sim A_{10}$	26.0	36.0
5	$A_{11} \sim A_{14}$	22.0	36.0
6	$A_{15} \sim A_{18}$	20.0	36.0
7	$A_{19} \sim A_{22}$	18.0	36.0

Table 6 Loading conditions for the 22-bar space truss

	Condition 1			Condition 2			Condition 3		
Node	P_X	P_Y	P_Z	P_X	P_Y	P_Z	P_X	P_Y	P_Z
1	-20.0	0.0	-5.0	-20.0	-5.0	0.0	-20.0	0.0	35.0
2	-20.0	0.0	-5.0	-20.0	-50.0	0.0	-20.0	0.0	0.0
3	-20.0	0.0	-30.0	-20.0	-5.0	0.0	-20.0	0.0	0.0
4	-20.0	0.0	-30.0	-20.0	-50.0	0.0	-20.0	0.0	-35.0

Note: loads are in kips

Table 7 Optimal design comparison for the 22-bar space truss

		Optimal cross-sectional areas (in.^{2})		
Variables		Sheu and Schmit [51]	Khan and Willmert [42]	***This work***
1	$A_1 \sim A_4$	2.629	2.563	***2.588***
2	$A_5 \sim A_6$	1.162	1.553	***1.083***
3	$A_7 \sim A_8$	0.343	0.281	***0.363***
4	$A_9 \sim A_{10}$	0.423	0.512	***0.422***
5	$A_{11} \sim A_{14}$	2.782	2.626	***2.827***
6	$A_{15} \sim A_{18}$	2.173	2.131	***2.055***
7	$A_{19} \sim A_{22}$	1.952	2.213	***2.044***
Weight (lb)		1024.80	1034.74	***1022.23***

Note: 1 in.^{2} = 6.452 cm^2, 1 lb = 4.45 N.

(5) Twenty-five-bar Space Truss

The 25-bar transmission tower space truss, shown in Figure 8, has been size optimized by many researchers. These include Schmit and Farshi [36], Schmit and Miura [37], Venkayya [38], Gellatly and Berke [39], Rizzi [41], Khan and Willmert [42], Templeman and Winterbottom [52], Chao et al. [53], Adeli and Kamal [54], John *et al.* [43], Saka [55], Fadel and Clitalay [56], Stander *et al.* [45], Xu and Grandhi [46], and Lamberti and Pappalettere [47,48]. In these studies, the material density was 0.1 lb/in.^{3} and modulus of elasticity was 10,000 ksi. This space truss was subjected to the two loading conditions shown in Table 8. The structure was required to be doubly symmetric about the *x*- and *y*-axes; this condition grouped the truss members as follows: (1) A_1, (2) $A_2 \sim A_5$, (3) $A_6 \sim A_9$, (4) $A_{10} \sim A_{11}$, (5) $A_{12} \sim A_{13}$, (6) $A_{14} \sim A_{17}$, (7) $A_{18} \sim A_{21}$, and (8) $A_{22} \sim A_{25}$.

The truss members were subjected to the compressive and tensile stress limitations shown in Table 9. In addition, maximum displacement limitations of ± 0.35 in. were imposed on every node in every direction. The minimum cross-sectional area of all members was 0.01 in.^{2}. The best HS algorithm solution vector for the eight design variables, obtained after approximately 15,000 searches, was (0.047, 2.022, 2.95, 0.01, 0.014, 0.688, 1.657, 2.663). The corresponding optimum weight was 544.38 lb. Table 10 gives a comparison between the optimal solutions reported in the literature and the present work. The HS algorithm produced a slightly better solution than any of the earlier mathematical studies.

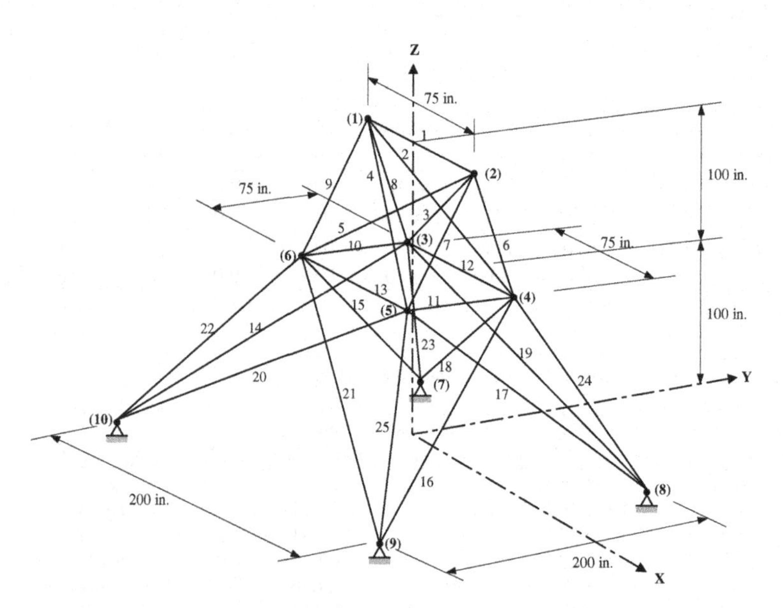

Fig. 8 Twenty-five-bar space truss.

Table 8 Loading conditions for the 25-bar space truss

	Condition 1			Condition 2		
Node	P_X	P_Y	P_Z	P_X	P_Y	P_Z
1	0.0	20.0	-5.0	1.0	10.0	-5.0
2	0.0	-20.0	-5.0	0.0	10.0	-5.0
3	0.0	0.0	0.0	0.5	0.0	0.0
6	0.0	0.0	0.0	0.5	0.0	0.0

Note: loads are in kips

Table 9 Member stress limitations for the 25-bar space truss

Variables		Compressive stress limitations (ksi)	Tensile stress limitations (ksi)
1	A_1	35.092	40.0
2	$A_2 \sim A_5$	11.590	40.0
3	$A_6 \sim A_9$	17.305	40.0
4	$A_{10} \sim A_{11}$	35.092	40.0
5	$A_{12} \sim A_{13}$	35.092	40.0
6	$A_{14} \sim A_{17}$	6.759	40.0
7	$A_{18} \sim A_{21}$	6.959	40.0
8	$A_{22} \sim A_{25}$	11.082	40.0

(6) Seventy-two-bar Space Truss

The 72-bar space truss, shown in Figure 9, has also been size optimized by many researchers, including Schmit and Farshi [36], Schmit and Miura [37], Venkayya [38], Gellatly and Berke [39], Khan and Willmert [42], Chao *et al.* [53], Adeli and Kamal [54], Berke and Khot [57], Xicheng and Guixu [58], Erbatur *et al.* [31], Adeli and Park [59], and Sarma and Adeli [32]. In these studies, the material density and modulus of elasticity were 0.1 lb/in.^{3} and 10,000 ksi, respectively. This space truss was subjected to the following two loading conditions: Condition 1, in which $P_X = 5.0$ kips, $P_Y = 5.0$ kips, and $P_Z = -5.0$ kips on node 17; and Condition 2, in which $P_X = 0.0$ kips, $P_Y = 0.0$ kips, and $P_Z = -5.0$ kips on nodes 17, 18, 19, and 20. The structure was required to be doubly symmetric about the *x*- and *y*-axes. This condition divided the truss members into the following sixteen groups: (1) $A_1 \sim A_4$, (2) $A_5 \sim A_{12}$, (3) $A_{13} \sim A_{16}$, (4) $A_{17} \sim A_{18}$, (5) $A_{19} \sim A_{22}$, (6) $A_{23} \sim A_{30}$, (7) $A_{31} \sim A_{34}$, (8) $A_{35} \sim A_{36}$, (9) $A_{37} \sim A_{40}$, (10) $A_{41} \sim A_{48}$, (11) $A_{49} \sim A_{52}$, (12) $A_{53} \sim A_{54}$, (13) $A_{55} \sim A_{58}$, (14) $A_{59} \sim A_{66}$, (15) $A_{67} \sim A_{70}$, and (16) $A_{71} \sim A_{72}$.

The members were subjected to stress limitations of ± 25 ksi, and the maximum displacement of uppermost nodes was not allowed to exceed ± 0.25 in. in the *x* and *y* directions. In this example, two cases were considered: Case1, in which the minimum cross-sectional area of all members was 0.1 in.^{2}; and Case2, in which the minimum cross-sectional area of 0.01 in.^{2} was considered.

Tables 11 and 12 show the HS algorithm's optimal results for Cases1 and 2 with the sixteen size variables, and compares these results with those previously reported in the literature. For Case1, the method proposed in this study achieved a design with the best solution vector of (1.790, 0.521, 0.100, 0.100, 1.229, 0.522, 0.100, 0.100, 0.517, 0.504, 0.100, 0.101, 0.156, 0.547, 0.442, 0.590) and a corresponding minimum weight of 379.27 lb after approximately 20,000 searches, which took ten minutes on a Pentium 600 MHz computer. For Case2, the best HS solution vector was (1.963, 0.481, 0.010, 0.011, 1.233, 0.506, 0.011, 0.012, 0.538, 0.533, 0.010, 0.167, 0.161, 0.542, 0.478, 0.551) and a corresponding minimum weight was 364.33 lb, which were also obtained after approximately 20,000 searches. The optimal design results for both cases obtained using the HS

Table 10 Comparison of optimal design for 25-bar space truss

		Optimal cross-sectional areas (in.^{2})								
Variables		Schmit and Farshi [36]	Schmit and Miura [37] NEWSUMT	Schmit and Miura [37] CONMIN	Venkayya [38]	Gellatly and Berke [39]	Rizzi [41]	Khan and Willmert [42]	Templeman and Winterbottom [52]	Chao et al. [53]
1	A_1	0.010	0.010	0.166	0.028	0.010	0.010	0.010	0.010	0.010
2	$A_2 \sim A_5$	1.964	1.985	2.017	1.964	2.007	1.988	1.755	2.022	2.042
3	$A_6 \sim A_9$	3.033	2.996	3.026	3.081	2.963	2.991	2.869	2.938	3.001
4	$A_{10} \sim A_{11}$	0.010	0.010	0.087	0.010	0.010	0.010	0.010	0.010	0.010
5	$A_{12} \sim A_{13}$	0.010	0.010	0.097	0.010	0.010	0.010	0.010	0.010	0.010
6	$A_{14} \sim A_{17}$	0.670	0.684	0.675	0.693	0.688	0.684	0.845	0.670	0.684
7	$A_{18} \sim A_{21}$	1.680	1.667	1.636	1.678	1.678	1.677	2.011	1.675	1.625
8	$A_{22} \sim A_{25}$	2.670	2.662	2.669	2.627	2.664	2.663	2.478	2.697	2.672
Weight (lb)		545.22	545.17	548.47	545.49	545.36	545.16	553.94	545.32	545.03

Note: 1 in.^{2} = 6.452 cm^2, 1 lb = 4.45 N. * Unavailable

Table 10 (*continued*)

Variables	Adeli and Kamal [54]	John et al. [43]	Saka [55]	Fadel and Clitalay [56]	Stander et al. [45]	Xu and Grandhi [46]	Lamberti and Pappalettere LEAML [47]	Lamberti and Pappalettere LESLP [48]	***This work***
1 A_1	0.010		0.010		0.010				***0.047***
2 $A_2 \sim A_5$	1.986		2.085		2.043				***2.022***
3 $A_6 \sim A_9$	2.961		2.988		3.003				***2.950***
4 $A_{10} \sim A_{11}$	0.010	*	0.010	*	0.010	*	*	*	***0.010***
5 $A_{12} \sim A_{13}$	0.010		0.010		0.010				***0.014***
6 $A_{14} \sim A_{17}$	0.806		0.696		0.683				***0.688***
7 $A_{18} \sim A_{21}$	1.680		1.670		1.623				***1.657***
8 $A_{22} \sim A_{25}$	2.530		2.592		2.672				***2.663***
Weight (lb)	545.66	555.58	545.23	545.49	545.03	545.60	545.17	545.17	***544.38***

Note: 1 in.^{2} = 6.452 cm^2, 1 lb = 4.45 N. * Unavailable

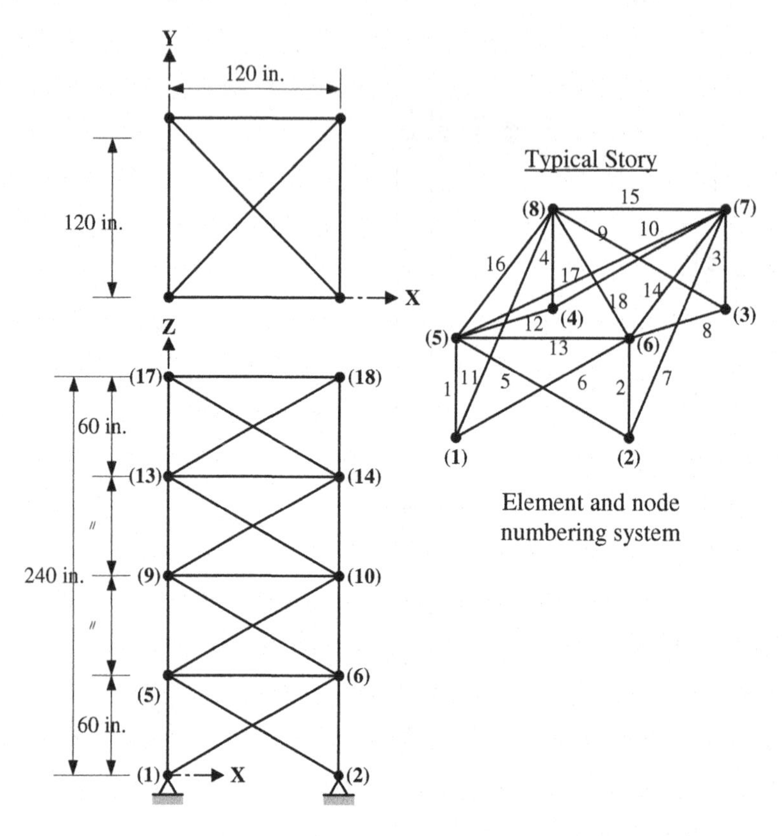

Fig. 9 Seventy-two-bar space truss.

approach were slightly better than all of the previous results using mathematical and genetic algorithms.

On the other hand, Figure 10 showed a comparison of convergence capability for Case 2 between the HS result and those obtained by Sarma and Adeli [32] using the simple and fuzzy genetic algorithm-based methods. A fuzzy genetic algorithm obtained a minimum weight of 364.40 lb after 1,758 structural analyses (B in Figure 10), while the HS algorithm obtained the same weight after 14,669 analyses (point b in Figure 10). The fuzzy controlled genetic algorithm method showed a better convergence capability than the present approach proposed on the basis of the pure HS algorithm. The proposed HS approach outperforms a simple genetic algorithm-based method in terms of both the convergence capability and the optimal solution, as shown in Figure 10. The simple genetic algorithm obtained a minimum weight of 372.40 lb after 2,776 analyses (A in Figure 10), but the HS approach required 1,076 structural analyses for the same weight (point a in Figure 10).

Table 11 Optimal design comparison for the 72-bar space truss (Case1)

	Optimal cross-sectional areas (in.^{2})						
		Schmit and Miura [37]				Khan and Willmert [42]	
Variables	Schmit and Farshi [36]	NEW-SUMT	CON-MIN	Venkayya [38]	Gellatly and Berke [39]	η=0.1	η=0.15
1 $A_1 \sim A_4$	2.078	1.885	1.885	1.818	1.464	1.793	1.859
2 $A_5 \sim A_{12}$	0.503	0.513	0.512	0.524	0.521	0.522	0.526
3 $A_{13} \sim A_{16}$	0.100	0.100	0.100	0.100	0.100	0.100	0.100
4 $A_{17} \sim A_{18}$	0.100	0.100	0.100	0.100	0.100	0.100	0.100
5 $A_{19} \sim A_{22}$	1.107	1.267	1.268	1.246	1.024	1.208	1.253
6 $A_{23} \sim A_{30}$	0.579	0.512	0.511	0.524	0.542	0.521	0.524
7 $A_{31} \sim A_{34}$	0.100	0.100	0.100	0.100	0.10	0.100	0.100
8 $A_{35} \sim A_{36}$	0.100	0.100	0.100	0.100	0.10	0.100	0.100
9 $A_{37} \sim A_{40}$	0.264	0.523	0.523	0.611	0.552	0.623	0.581
10 $A_{41} \sim A_{48}$	0.548	0.517	0.5161	0.532	0.608	0.523	0.527
11 $A_{49} \sim A_{52}$	0.100	0.100	0.100	0.100	0.100	0.100	0.100
12 $A_{53} \sim A_{54}$	0.151	0.100	0.113	0.100	0.100	0.196	0.158
13 $A_{55} \sim A_{58}$	0.158	0.157	0.156	0.161	0.149	0.149	0.152
14 $A_{59} \sim A_{66}$	0.594	0.546	0.548	0.557	0.773	0.570	0.561
15 $A_{67} \sim A_{70}$	0.341	0.411	0.411	0.377	0.453	0.443	0.438
16 $A_{71} \sim A_{72}$	0.608	0.570	0.561	0.506	0.342	0.519	0.532
Weight (lb)	388.63	379.64	379.79	381.2	395.97	381.72	387.67

Note: 1 in.^{2} = 6.452 cm^2, 1 lb = 4.45 N.

Table 11 *(continued)*

	Optimal cross-sectional areas (in.^{2})						
					Erbatur *et al.* [31]		
Variables	Chao et al. [53]	Adeli and Kamal [54]	Berke and Khot [57]	Xicheng and Guixu [58]	GAOS1	GAOS2	***This work***
1 $A_1 \sim A_4$	1.832	2.026	1.893	1.905	1.755	1.910	***1.790***
2 $A_5 \sim A_{12}$	0.512	0.533	0.517	0.518	0.505	0.525	***0.521***
3 $A_{13} \sim A_{16}$	0.100	0.100	0.100	0.100	0.105	0.122	***0.100***
4 $A_{17} \sim A_{18}$	0.100	0.100	0.100	0.100	0.155	0.103	***0.100***
5 $A_{19} \sim A_{22}$	1.252	1.157	1.279	1.286	1.155	1.310	***1.229***
6 $A_{23} \sim A_{30}$	0.524	0.569	0.515	0.516	0.585	0.498	***0.522***
7 $A_{31} \sim A_{34}$	0.100	0.100	0.100	0.100	0.100	0.110	***0.100***
8 $A_{35} \sim A_{36}$	0.100	0.100	0.100	0.100	0.100	0.103	***0.100***
9 $A_{37} \sim A_{40}$	0.513	0.514	0.508	0.509	0.460	0.535	***0.517***
10 $A_{41} \sim A_{48}$	0.529	0.479	0.520	0.522	0.530	0.535	***0.504***
11 $A_{49} \sim A_{52}$	0.100	0.100	0.100	0.100	0.120	0.103	***0.100***
12 $A_{53} \sim A_{54}$	0.100	0.100	0.100	0.100	0.165	0.111	***0.101***
13 $A_{55} \sim A_{58}$	0.157	0.158	0.157	0.157	0.155	0.161	***0.156***
14 $A_{59} \sim A_{66}$	0.549	0.550	0.539	0.537	0.535	0.544	***0.547***
15 $A_{67} \sim A_{70}$	0.406	0.345	0.416	0.411	0.480	0.379	***0.442***
16 $A_{71} \sim A_{72}$	0.555	0.498	0.551	0.571	0.520	0.521	***0.590***
Weight (lb)	379.62	379.31	379.67	380.84	385.76	383.12	***379.27***

Note: 1 in.^{2} = 6.452 cm^2, 1 lb = 4.45 N.

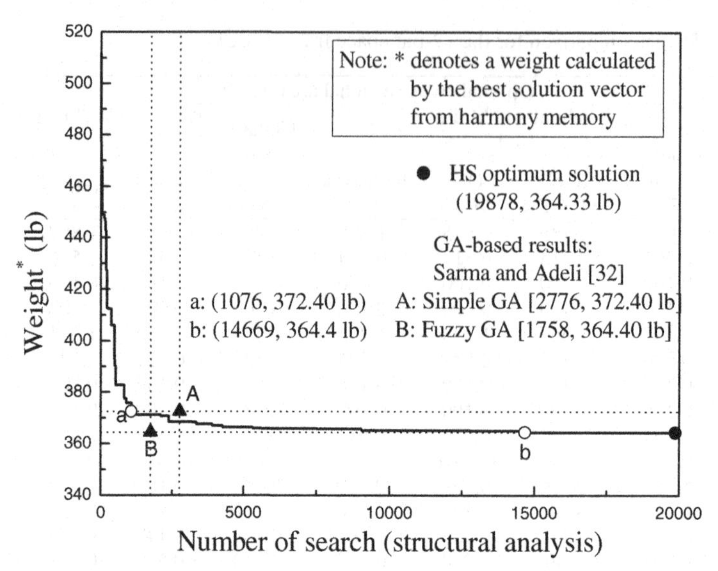

Fig. 10 Convergence history of the minimum weight for 72-bar space truss (Case2).

(7) Two-hundred-bar Planar Truss

The 200-bar plane truss, shown in Figure 11, has been size optimized using mathematical methods by Stander *et al.* [45] and Lamberti and Pappalettere [47, 48]. All members are made of steel: the material density and modulus of elasticity were 0.283 lb/in.^{3} and 30,000 ksi, respectively. This truss was subjected to constraints only on stress limitations of ± 10 ksi. There were three loading conditions: (1) 1.0 kip acting in the positive x-direction at nodes 1, 6, 15, 20, 29, 34, 43, 48, 57, 62, and 71; (2) 10 kips acting in the negative y-direction at nodes 1, 2, 3, 4, 5, 6, 8, 10, 12, 14, 15, 16, 17, 18, 19, 20, 22, 24,…, 71, 72, 73, 74, and 75; and (3) conditions 1 and 2 acting together.

The 200 members of this truss were linked into twenty-nine groups, as shown in Table 13. The minimum cross-sectional area of all members was 0.1 in.^{2} The HS algorithm-based method was applied to the 200-bar truss with twenty-nine independent design variables.

It found 20 different solution vectors after 50,000 searches. The best result is compared to the earlier solutions reported by Stander *et al.* [45] and Lamberti and Pappalettere [47, 48] in Table 13. The HS algorithm-based method found an optimum weight of 25,447.1 lb after approximately 48,000 searches. The optimal design obtained using the HS algorithm showed an excellent agreement with the previous mathematical designs reported in the literature.

(8) One-hundred-twenty-bar Dome Truss

The design of 120-bar dome truss, shown in Figure 12, was considered as a last example to demonstrate the practical capability of the HS heuristic algorithm-based method. This dome truss was first analyzed by Soh and Yang [28] to obtain the optimal sizing and configuration variables (*i.e.*, the structural configuration

Table 12 Optimal design comparison for the 72-bar space truss (Case2)

	Variables	Optimal cross-sectional areas (in.^{2})			
		Adeli and Park [59]	Sarma and Adeli [32]		
			Simple GA	Fuzzy GA	*This work*
1	$A_1 \sim A_4$	2.755	2.141	1.732	***1.963***
2	$A_5 \sim A_{12}$	0.510	0.510	0.522	***0.481***
3	$A_{13} \sim A_{16}$	0.010	0.054	0.010	***0.010***
4	$A_{17} \sim A_{18}$	0.010	0.010	0.013	***0.011***
5	$A_{19} \sim A_{22}$	1.370	1.489	1.345	***1.233***
6	$A_{23} \sim A_{30}$	0.507	0.551	0.551	***0.506***
7	$A_{31} \sim A_{34}$	0.010	0.057	0.010	***0.011***
8	$A_{35} \sim A_{36}$	0.010	0.013	0.013	***0.012***
9	$A_{37} \sim A_{40}$	0.481	0.565	0.492	***0.538***
10	$A_{41} \sim A_{48}$	0.508	0.527	0.545	***0.533***
11	$A_{49} \sim A_{52}$	0.010	0.010	0.066	***0.010***
12	$A_{53} \sim A_{54}$	0.643	0.066	0.013	***0.167***
13	$A_{55} \sim A_{58}$	0.215	0.174	0.178	***0.161***
14	$A_{59} \sim A_{66}$	0.518	0.425	0.524	***0.542***
15	$A_{67} \sim A_{70}$	0.419	0.437	0.396	***0.478***
16	$A_{71} \sim A_{72}$	0.504	0.641	0.595	***0.551***
Weight (lb)		376.50	372.40	364.40	***364.33 [364.40]*1 [372.40]*2***
Number of structural analyses		-	2776	1758	***19878 [14669]*1 [1076]*2***

Note: 1 in.^{2} = 6.452 cm^2, 1 lb = 4.45 N.

*1 HS obtained a weight of 364.40 lb after 14669 analyses (the result of Fuzzy GA)

*2 HS obtained a weight of 372.40 lb after 1076 analyses (the result of Simple GA)

optimization). In the example considered in this study, however, only sizing variables to minimize the structural weight were considered. In addition, the allowable tensile and compressive stresses were used according to the AISC ASD (1989) [60] code, as follows:

$$\begin{aligned} {}_U\sigma_i &= 0.6F_y \quad for \quad \sigma_i \geq 0 \ (Tensile \quad stress) \\ {}_L\sigma_i & \quad for \quad \sigma_i < 0 \ (Compressive \quad stress) \end{aligned} \tag{8}$$

$${}_L\sigma_i = \begin{cases} \left[\left(1-\dfrac{\lambda_i^2}{2C_c^2}\right)F_y\right] \Big/ \left(\dfrac{5}{3}+\dfrac{3\lambda_i}{8C_c}-\dfrac{\lambda_i^3}{8C_c^3}\right) & for\, \lambda_i < C_c \\ \dfrac{12\pi^2 E}{23\lambda_i^2} \quad for\, \lambda_i \geq C_c \end{cases} \tag{9}$$

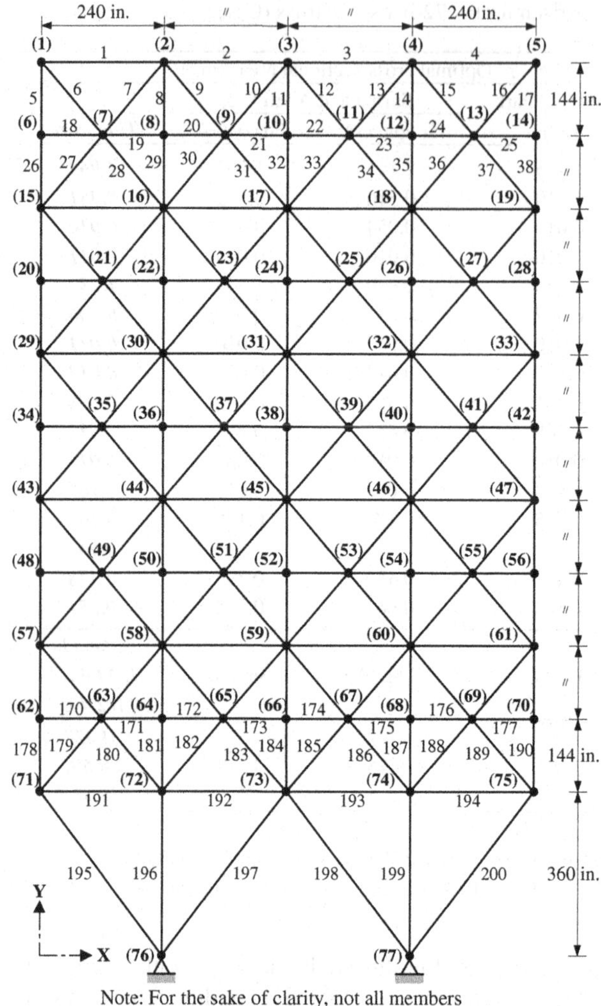

Fig. 11 Two-hundred-bar planar truss.

where E = the modulus of elasticity; F_y = the yield stress of steel; C_c = the slenderness ratio (λ_i) dividing the elastic and inelastic buckling regions ($C_c = \sqrt{2\pi^2 E / F_y}$); λ_i = the slenderness ratio ($\lambda_i = kL_i / r_i$); k = the effective length factor; L_i = the member length; and r_i = the radius of gyration.

The modulus of elasticity (E) was 30,450 ksi and the material density was 0.288 lb/in.^{3}. The yield stress of steel (F_y) was taken as 58.0 ksi. The radius of gyration (r_i) can be expressed in terms of cross-sectional areas, *i.e.*, $r_i = aA_i^b$ [55]. Here, a and b are the constants depending on the types of sections adopted for the

Table 13 Optimal design comparison for the 200-bar planar truss

Group	Variables Members (A_i, i=1,200)	Optimal cross-sectional areas (in.^{2})	Comparison Optimizer	 Weight (lb)
1	1,2,3,4	0.1253	This work	*25447.1*
2	5,8,11,14,17	1.0157		
3	19,20,21,22,23,24	0.1069		
4	18,25,56,63,94,101,132,139,170,177	0.1096	Stander *et al.* [45]	25446.70
5	26,29,32,35,38	1.9369		
6	6,7,9,10,12,13,15,16,27,28,30,31,33	0.2686		
	34,36,37		Lamberti	25446.2 (LEAML)
7	39,40,41,42	0.1042	And	25450.3 (CGML1)
8	43,46,49,52,55	2.9731	Pappalettere [47]	25446.9 (CGML2)
9	57,58,59,60,61,62	0.1309		25450.4 (CGML3)
10	64,67,70,73,76	4.1831		25450.2 (CGML4)
11	44,45,47,48,50,51,53,54,65,66,68,69	0.3967		25450.2 (CGML5)
	71,72,74,75			25446.7 (CGML6)
12	77,78,79,80	0.4416		
13	81,84,87,90,93	5.1873		
14	95,96,97,98,99,100	0.1912	Lamberti and Pappalettere	25446.17 (LESLP)
15	102,105,108,111,114	6.2410	[48]	25446.17 (DOT)
16	82,83,85,86,88,89,91,92,103,104,106	0.6994		
	107,109,110,112,113			
17	115,116,117,118	0.1158		
18	119,122,125,128,131	7.7643		
19	133,134,135,136,137,138	0.1000		
20	140,143,146,149,152	8.8279		
21	120,121,123,124,126,127,129,130,141	0.6986		
	142,144,145,147,148,150,151			
22	153,154,155,156	1.5563		
23	157,160,163,166,169	10.9806		
24	171,172,173,174,175,176	0.1317		
25	178,181,184,187,190	12.1492		
26	158,159,161,162,164,165,167,168,179	1.6373		
	180,182,183,185,186,188,189			
27	191,192,193,194	5.0032		
28	195,197,198,200	9.3545		
29	196,199	15.0919		

Note: 1 in.^{2} = 6.452 cm^2, 1 lb = 4.45 N.

members such as pipes, angles, and tees. In this example, pipe sections (a = 0.4993 and b = 0.6777) were adopted for bars. All members of the dome were linked into seven groups, as shown in Figure 12.

The dome was considered to be subjected to vertical loading at all the unsupported joints. These were taken as -13.49 kips at node 1, -6.744 kips at nodes 2 through 13, and -2.248 kips at the rest of the nodes. The minimum cross-sectional area of all members was 0.775 in.^{2} . In this example, two cases of displacement constraints were considered: no displacement constraints (Case1) and displacement limitations of ± 0.1969 in. imposed on all nodes in x- and y-directions.

Table 14 gives the best solution vectors and the corresponding weights for Cases 1 and 2, respectively. Both design procedures obtained each optimum solution after approximately 35,000 searches.

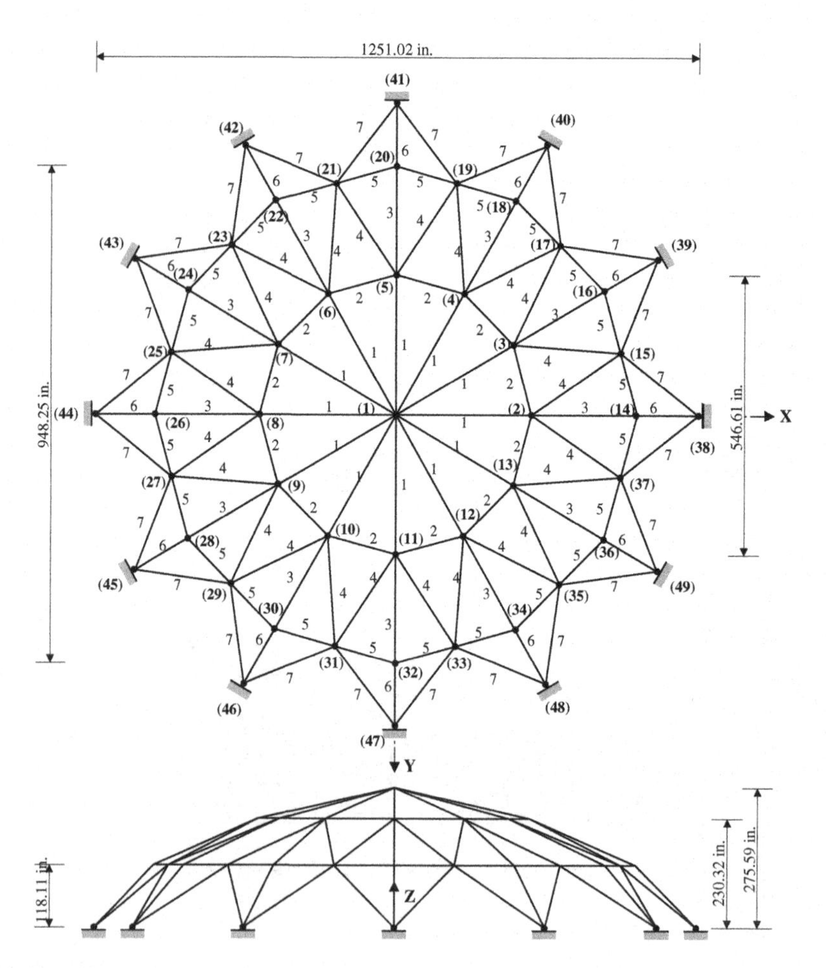

Fig. 12 One-hundred-twenty-bar dome truss.

Table 14 Optimal design for the 120-bar dome truss

Variables (groups)	Optimal cross-sectional areas (in.^{2})	
	Case1	Case2
1	3.295	3.296
2	2.396	2.789
3	3.874	3.872
4	2.571	2.570
5	1.150	1.149
6	3.331	3.331
7	2.784	2.781
Weight (lb)	19707.77	19893.34

Note: 1 in.^{2} = 6.452 cm^2, 1 lb = 4.45 N.

4.2 Continuous Configuration Optimization Examples

Three classical truss examples were presented to demonstrate the search efficiency of the continuous configuration optimization approach using BHS algorithm. Values for the BHS algorithm parameters (*i.e.*, HMS, HMCR, and PAR) used in all examples were arbitrarily selected, based on those recommended by Geem [34], as previously stated. On the other hand, the following strategy recommended by Geem [34] was adopted for generating the initial HM (solution vectors) for all examples (*i.e.*, initialization step): (1) first, 200 different feasible solution vectors were randomly generated from all possible variable bounds based on the FEM structural analysis method subject to the constraint functions; (2) better solution vectors judged in terms of objective function values among the generated 200 feasible solutions were then selected as many as the size of the HM (*i.e.*, HMS) for the initial HM of each example.

(1) Ten-bar Plane Truss

The 10-bar cantilever plane truss structure shown in Figure 4 is one of the most popular classical optimization design problems. Due to its simple configuration, the 10-bar truss has been used as a benchmark to verify the efficiency of various optimization methods. The material density of this truss was 7.875 g/cm^3 (0.283 lb/in.^{3}) and the modulus of elasticity was 206.7 GPa (30000 ksi). Displacement limits of 5.08 cm (2.0 in.) were imposed on node 2 in both directions, and the limiting tensile and compressive stresses in each member were 175.25 MPa (25 ksi). The HS algorithm parameters were as follows: the HMS was 10, the HMCR was 0.95, the PAR was 0.3, and the maximum number of searches was 50000.

No cross-sectional area variable linking was used, and the upper nodes, 1 and 3, were allowed to move in the vertical direction (Y). Thus, there were twelve independents design variables that include ten sizing variables and two coordinate variables. The bounds on the member cross-sectional areas were 0.452 - 64.52 cm^2 (0.07 - 10 in^2). The bounds on the node coordinates were 508 - 1270 cm (200 - 500 in) for both Y_1 and Y_3. In this example, two cases were investigated: Case 1, in which only the stress constraints were considered; and Case 2, in which both the stress and displacement constraints were considered. The single load condition, *i.e.*, P_1 and P_2 = 444.5 kN (100 kips), was considered in each case.

The HS algorithm was applied to each case and the optimal results were compared to earlier solutions reported by Yang and Soh [19], as shown in Table 15. Yang and Soh solved the problem using a pure GA-based approach, obtaining a weight of 19887.5 N (4471.1 lb) after 2840 structural analyses for Case 1 and 22707.0 N (5105.0 lb) after 2080 analyses for Case 2, respectively. For Case 1, the HS algorithm found a minimum weight of 19747.8 N (4439.7 lb) after 2655 searches (FEM structural analyses), which was better optimized than the value obtained by Yang and Soh. Note that the number of structural analyses used in the present approach is also less than the method used by Yang and Soh. For Case 2, the HS algorithm found a minimum weight of 22291.6 N (5011.6 lb) after 5861 searches, as shown in Table 15, which was also better optimized than the value obtained by Yang and Soh. Figure 13 showed a comparison of convergence capability for Case 2 between the HS result and that obtained by Yang and Soh. It is noteworthy to mention from Figure 13 that Yang and Soh obtained a minimum weight of 22707.0 N (5105.0 lb) after 2080 structural analyses using the pure GA approach, while the proposed HS approach obtained a weight of 22300.9 N (5013.7 lb) (point a in Figure 13) at the same number of analyses.

Table 15 Optimal Result for 10-Bar Plane Truss

Design variables A_i (cm^2) and R_i (cm)	Case 1		Case 2	
	Yang & Soh[19]	*This Work*	Yang & Soh[19]	*This work*
A_1	51.48	***51.34***	59.35	***62.40***
A_2	0.710	***0.961***	0.581	***0.548***
A_3	53.87	***54.32***	54.45	***56.16***
A_4	27.23	***27.28***	30.26	***38.23***
A_5	1.161	***0.458***	0.484	***0.490***
A_6	0.645	***0.839***	1.161	***0.452***
A_7	37.10	***37.03***	37.10	***36.14***
A_8	32.32	***31.23***	40.97	***33.34***
A_9	37.05	***36.99***	49.23	***48.31***
A_{10}	0.774	***1.071***	1.290	***0.574***
X_1	1828.8[a]	***1828.8[a]***	1828.8[a]	***1828.8[a]***
Y_1	609.6	***508.26***	763.3	***511.13***
X_3	914.4[a]	***914.4[a]***	914.4[a]	***914.4[a]***
Y_3	851.7	***837.85***	846.6	***771.67***
Weight (N)	19887.5	***19747.8***	22707.0	***22291.6 [22300.9][b]***
Number of structural analyses	2840	***2655***	2080	***5861 [2080][b]***

[a]Coordinate is stationary. [b]HS obtained a weight of 22300.9 N after 2080 analyses.

(2) Eighteen-bar Plane Truss

Figure 6 showed the initial configuration of an 18-bar cantilever plane steel truss, which was previously solved by Imai and Schmit [50], Felix [61], Yang [62], Soh and Yang [28], Rajeev and Krishnamoorthy [22], and Yang and Soh [63]. The cross-sectional areas of the members have been categorized into four groups, as follows: (1) $A_1=A_4=A_8=A_{12}=A_{16}$, (2) $A_2=A_6=A_{10}=A_{14}=A_{18}$, (3) $A_3=A_7=A_{11}=A_{15}$, and (4) $A_5=A_9=A_{13}=A_{17}$. The single loading condition was a set of vertical loads, P = 88.96 kN (20 kips), acting on the upper nodal points of the truss, as illustrated

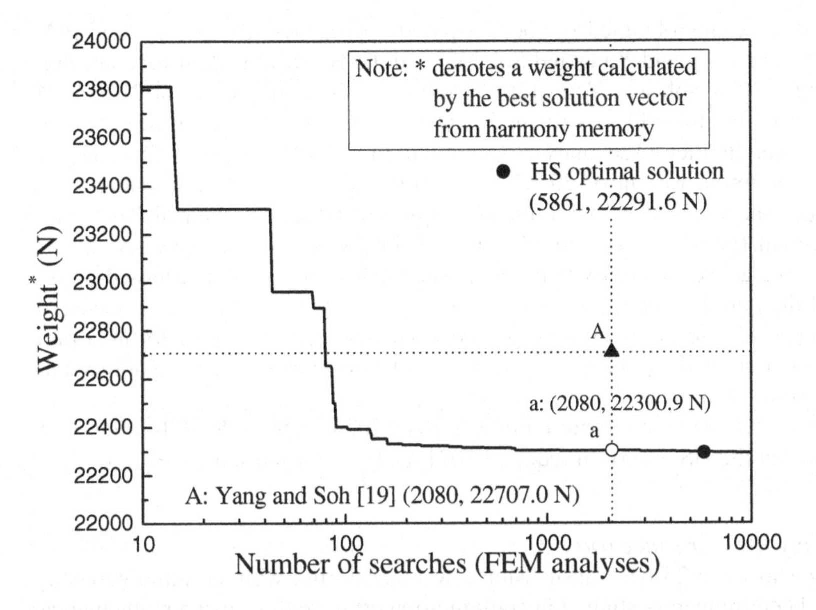

Fig. 13 Convergence History of the Minimum Weight for 10-Bar Plane Truss (Case 2)

in Figure 6. The lower nodes, 3, 5, 7, and 9, were allowed to move in any direction in the X-Y plane. Thus, there were a total of twelve independent design variables that include four sizing and eight coordinate variables. The purpose of the optimization is to design a configuration for the truss to produce a minimum design weight that meets both the allowable stress and the buckling constrains. The Euler buckling compressive stress limit for truss member i was used for the buckling constrains. It was computed as

$$_{b}\sigma_i = -KEA_i / L_i^2 \tag{10}$$

where K = a constant determined from the cross-sectional geometry, E = modulus of elasticity of the material, and L_i = the member length. In this study, the buckling constant was taken to be $K = 4$. The other design data were as follows: modulus of elasticity was 68.9 GPa (10000 ksi), material density was 2.768 g/cm^3 (0.1 lb/in.^{3}), and the allowable tensile and compressive stresses were 137.9 MPa (20 ksi). The bounds on the member cross-sectional areas were 22.58-116.13 cm^2 (3.5-18.0 in.^{2}). The HS algorithm parameters were HMS = 20, HMCR = 0.90, PAR = 0.4, and the maximum number of searches = 50000. Table 16 presents the best solution vector from the HS and also the results obtained using other mathematical methods (Imai and Schmit [50] and Felix [61]) and GA-based approaches (Yang [62], Soh and Yang [28], Rajeev and Krishnamoorthy [22], and Yang and Soh [63]). The HS algorithm found a minimum weight of 20085.4 N (4515.6 lb) after 24805 searches, which was better optimized than the other six results reported in the literature. Figure 14 showed a comparison of convergence capability between the results obtained by the GA-based approaches studied by Yang [62], Soh and Yang [28], and Yang and Soh [63] and the HS results. A fuzzy controlled GA proposed by Soh and Yang [28] obtained a minimum weight of 20157.9 N (4531.9 lb) after 1440 FEM structural analyses, while the HS algorithm obtained the same weight after 3426 analyses (point b in Figure 14). Yang and Soh [63] obtained a minimum weight of 20105 N (4520 lb) after 1200 analyses using a GA with tournament selection, while the HS approach required 9506 analyses for the same weight (point a in Figure 14). Both hybrid GA-based approaches show a better convergence capability than the present approach that was proposed on the basis of the pure HS algorithm with the rejecting strategy for the fitness measure. However, it should be noted that the proposed HS approach outperforms a pure GA method studied by Yang [62] in terms of both convergence capability and optimal solution.

The pure GA obtained a minimum weight of 20250.9 N (4552.8 lb) after 3000 analyses, but the HS approach required 2071 analyses for the same weight (point c in Figure 14).

(3) Twenty five-bar space truss

Figure 8 showed a 25-bar transmission tower space truss with an initial geometry that has been frequently studied in configuration optimizations using mathematical approaches (Vanderplaats and Moses [64]; Felix [61]; Hansen and Vanderplaats [65]) and GA-based approaches (Yang [62]; Soh and Yang [28]; Yang and Soh [63]). The tower was subjected to two loading conditions shown in Table 8. The material density of this truss was 2.768 g/cm^3 (0.1 lb/in.^{3}) and modulus of elasticity was 68.9 GPa (10000 ksi). All members were constrained to 275.6 MPa (40 ksi) in both tension and compression. In addition, all member stresses were constrained to the Euler buckling stress, as given by Eq. (7), with the buckling constant K = 39.274 corresponding to tubular members with a nominal diameter-to-thickness ratio of 100.

The HS algorithm parameters were as follows: HMS = 20, HMCR = 0.8, PAR = 0.4, and the maximum number of searches = 50000. The geometric variables were selected as coordinates X_4, Y_4, Z_4, X_8, and Y_8 with symmetry required in the X-Z and Y-Z planes. The cross sectional-areas of the members were linked into the eight groups shown in Figure. 8. Thus, there were a total of thirteen design variables that included eight sizing variables and five independent coordinate variables for two loading conditions. The lower and upper bounds on the member cross-sectional areas were 0.065 - 6.45 cm^2 (0.01 - 1.0 in^2).

The HS algorithm-based approach was applied to the space truss. Table 17 presents the optimal results along with those reported by Vanderplaats and Moses [64], Felix [61], and Hansen and Vanderplaats [64] using mathematical methods and Yang [62], Soh and Yang [28], and Yang and Soh [63] using GA-based approaches. After 39068 searches (FEM structural analyses), the best solution vector for the thirteen design variables and corresponding objective function value (weight of the structure) were obtained from the HS, as shown in the table. The HS found a minimum weight of 566.2 N (127.3 lb), which is better than the values obtained in all of the previous investigations.

Table 16 Optimal Result for 18-Bar Plane Truss

Design variables A_i (cm^2) and R_i (cm)	Imai & Schmit [50]	Felix [61]	Yang [62]	Soh & Yang [28]	Rajeev & Krishnamoorthy [22]	Yang & Soh [63]	***This Work***
$A_1=A_4=A_8=A_{12}=A_{16}$	72.55	73.16	81.36	81.23	80.65	79.55	***81.59***
$A_2=A_6=A_{10}=A_{14}=A_{18}$	101.17	124.39	116.77	115.56	104.84	115.94	***111.10***
$A_3=A_7=A_{11}=A_{15}$	51.19	70.79	35.29	35.48	51.61	36.13	***39.82***
$A_5=A_9=A_{13}=A_{17}$	41.90	34.19	22.84	22.92	25.81	23.61	***22.88***
X_3	2263.4	2526.3	2322.8	2310.9	2265.4	2304.3	***2293.8***
Y_3	364.7	412.2	464.9	468.6	369.0	467.9	***442.8***
X_5	1544.8	1898.4	1643.3	1626.4	1550.9	1634.0	***1601.0***
Y_5	267.7	261.4	374.3	375.5	300.2	379.0	***346.1***
X_7	969.5	1226.6	1052.1	1041.5	978.9	1051.4	***1021.3***
Y_7	145.0	083.8	254.9	246.4	184.2	259.1	***229.9***
X_9	459.7	563.1	508.1	510.2	468.4	513.3	***496.0***
Y_9	-8.1	43.4	81.0	81.2	59.4	78.5	***77.8***
Weight (N)	20762.8	25411.4	20250.9	20157.9	20535.5	20105.0	***20085.4 [20105.0]***[a] ***[20157.9]***[b] ***[20250.9]***[c]
Number of structural analyses	111	78	3000	1440	-	1200	***24805 [9506]***[a] ***[3426]***[b] ***[2071]***[c]

[a]HS obtained a weight of 20105.0 N after 9506 analyses (the result of Yang and Soh [63]).
[b]HS obtained a weight of 20157.9 N after 3426 analyses (the result of Soh and Yang [28]).
[c]HS obtained a weight of 20250.9 N after 2071 analyses (the result of Yang [62]).

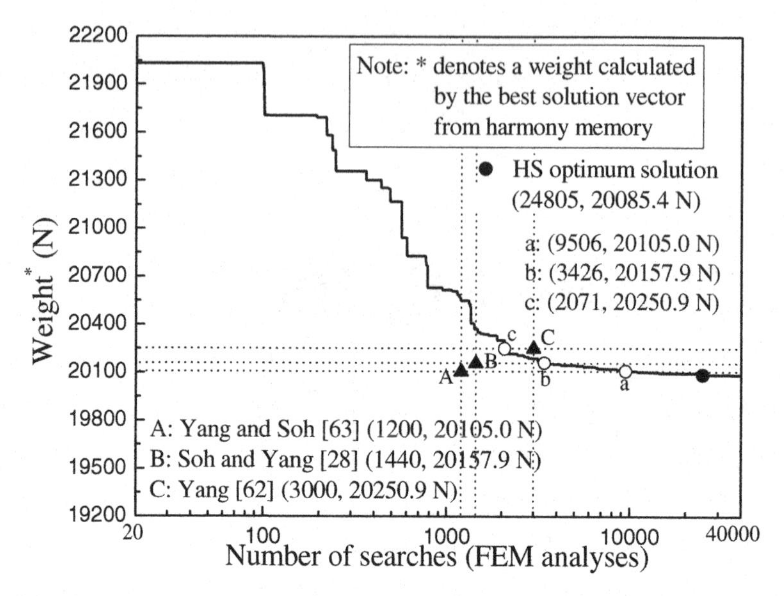

Fig. 14 Convergence History of the Minimum Weight for 18-Bar Plane Truss

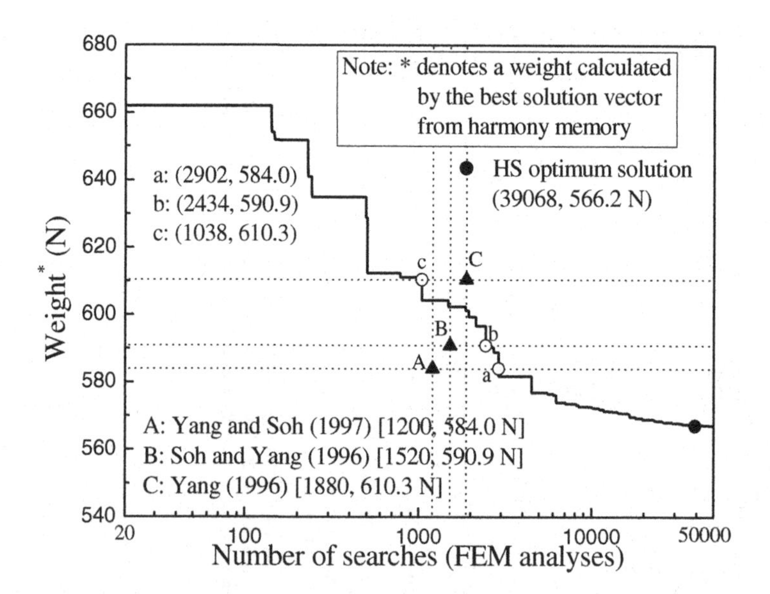

Fig. 15 Convergence History of the Minimum Weight for 25-Bar Space Truss

Figure 15 showed a comparison of convergence capability between the HS results and those obtained by the GA-based approaches. A fuzzy controlled GA (Soh and Yang [28]) obtained a minimum weight of 590.9 N (132.8 lb) after 1520 structural analyses, while 2434 analyses for the HS was required to obtain the same weight (point b in Figure 15). A GA with tournament selection (Yang and Soh [63]) obtained a minimum weight of 584 N (131.3 lb) after 1200 analyses, while the HS obtained the same weight after 2902 analyses (point a in Figure 15). On the other hand, a pure GA method (Yang [62]) optimized a minimum weight of 610.3 N (137.2 lb) after 1880 analyses, but 1038 analyses for the HS approach was required to reach the same weight (point c in Figure 15). Hybrid GA-based methods proposed by Soh and Yang [28] and Yang and Soh [63] show a better convergence capability than the HS approach, but the convergence capability of the HS outperforms the pure GA (Yang [62]), as shown in the Figure 15. These results are similar to those shown in Figure 14 (18-bar plane truss).

Table 17 Optimal Result for 25-Bar Space Truss

Design variables A_i (cm^2) & R_i (cm)	Vanderplaats & Moses [64]	Felix [61]	Hansen & Vanderplaats [64]	Yang [62]	Soh & Yang [28]	Yang & Soh [63]	***This work***
A_1	0.21	0.08	0.07		0.58		***0.14***
A_2	3.65	2.67	3.14		2.84		***2.35***
A_3	5.23	5.43	5.39		5.81		***5.62***
A_4	0.18	0.21	0.16	-	0.32	-	***0.25***
A_5	0.30	0.65	0.79		0.71		***0.62***
A_6	0.63	0.78	0.54		1.36		***0.79***
A_7	4.83	4.77	4.50		4.52		***5.34***
A_8	3.55	3.57	3.54		3.61		***3.70***
X_4	32.8	54.6	60.2	57.5	55.8	57.1	***51.4***
Y_4	122.4	122.7	125.2	106.7	110.7	124.2	***95.9***
Z_4	247.4	254.8	248.2	251.1	246.0	255.5	***262.9***
X_8	94.2	56.1	69.9	39.6	35.9	64.0	***45.4***
Y_8	239.0	244.7	244.9	209.2	206.1	249.3	***198.8***
Weight (N)	593.8	571.6	570.7	610.3	590.9	584.0	***566.2 [584.0][a] [590.9][b] [610.3][c]***
Number of structural analyses	171	-	7	1880	1520	1200	***39068 [2902][a] [2434][b] [1038][c]***

[a]HS obtained a weight of 584.0 N after 2902 analyses (the result of Yang and Soh [63]).
[b]HS obtained a weight of 590.9 N after 2434 analyses (the result of Soh and Yang [28]).
[c]HS obtained a weight of 610.3 N after 1038 analyses (the result of Yang [62]).

4.3 Discrete Size Optimization Examples

The previously described computational procedures were implemented in a FORTRAN computer program that was applied to discrete sizing configuration optimization problems for trusses. The FEM displacement method was used to analyze the truss structures.

Standard test truss examples were considered to demonstrate the discrete search efficiency of the HS algorithm approach, as compared to current methods. The cases shown in Table 18, each with a different set of HS algorithm parameters (*i.e.*, HMS, HMCR, and PAR), were tested with all of the examples presented in this study. These parameter values were arbitrarily selected, based on the empirical findings by Geem [34], which determined that the HS algorithm performed well with $10 \leq \text{HMS} \leq 50$, $0.7 \leq \text{HMCR} \leq 0.95$, and $0.2 \leq \text{PAR} \leq 0.5$. The maximum number of searches was set to 30,000.

Table 18 HS Algorithm Parameters Used for All Examples

Cases	HMS	HMCR	PAR
Case-1	20	0.9	0.45
Case-2	40	0.9	0.45
Case-3	30	0.9	0.4
Case-4	30	0.8	0.3
Case-5	30	0.9	0.3

(1)Twenty Five-bar Transmission Tower Space Truss

The 25-bar transmission tower space truss, shown in Figure 8, has been optimized using discrete size algorithms by many researchers, including Rajeev and Krishnamoorthy [21]), Wu and Chow [26, 27], Adeli and Park [66], Erbatur *et al.* [31], and Park and Sung [67]. In these studies, the material density was 0.1 lb/in.^{3} and modulus of elasticity was 10,000 ksi. This space truss was subjected to the following loading condition: $P_X = 1.0$ kips and $P_Y = P_Z = -10.0$ kips acting on node 1, $P_X = 0.0$ kips and $P_Y = P_Z = -10.0$ kips acting on node 2, $P_X = 0.5$ kips and $P_Y = P_Z = 0.0$ kips acting on node 3, and $P_X = 0.6$ kips and $P_Y = P_Z = 0.0$ kips acting on node 6. The structure was required to be doubly symmetric about the *X*- and *Y*-axes; this condition grouped the truss members as follows: (1) A_1, (2) $A_2 \sim A_5$, (3) $A_6 \sim A_9$, (4) $A_{10} \sim A_{11}$, (5) $A_{12} \sim A_{13}$, (6) $A_{14} \sim A_{17}$, (7) $A_{18} \sim A_{21}$, and (8) $A_{22} \sim A_{25}$. All members were constrained to 40 ksi in both tension and compression. In addition, maximum displacement limitations of ± 0.35 in. were imposed at each node in every direction. Discrete values for the cross-sectional areas were taken from the set $D \in \{0.1, 0.2, 0.3, 0.4, 0.5, 0.6, 0.7, 0.8, 0.9, 1.0, 1.1, 1.2, 1.3, 1.4, 1.5, 1.6, 1.7, 1.8, 1.9, 2.0, 2.1, 2.2, 2.3, 2.4, 2.5, 2.6, 2.8, 3.0, 3.2, 3.4\}$ (in.^{2}), which has thirty discrete values.

The HS algorithm-based discrete size optimization approach was applied to the space truss. Table 19 lists the HS result obtained with each set of parameters given in Table 16. The results reported by Rajeev and Krishnamoorthy [21], Wu and Chow [26, 27], and Erbatur *et al.* [31], obtained with GA-based methods, by Adeli and Park [66], obtained with the neural dynamics model, and by Park and Sung [67], obtained with the simulated annealing algorithm-based method, are also included in the table. After 13,523 to 18,734 searches (FEM structural analyses), the best solution vector and the corresponding objective function value (the structural

Table 19 Optimal Results of 25-Bar Space Truss

Design variables A_i (in.^{2})		*HS results* Case-1	Case-2	Case-3	Case-4	Case-5	Rajeev *et al.* [21]	Wu & Chow [26]	Wu & Chow [27]	Adeli & Park [66]	Erbatur *et al.* [31]	Park & Sung [97]
1	A_1	***0.1***	***0.1***	***0.1***	***0.1***	***0.1***	0.1	0.1	0.1	0.6	0.1	0.1
2	$A_2 \sim A_5$	***0.6***	***0.3***	***0.3***	***0.5***	***0.3***	1.8	0.6	0.5	1.4	1.2	2.1
3	$A_6 \sim A_9$	***3.4***	***3.4***	***3.4***	***3.4***	***3.4***	2.3	3.2	3.4	2.8	3.2	3.4
4	$A_{10} \sim A_{11}$	***0.1***	***0.1***	***0.1***	***0.1***	***0.1***	0.2	0.2	0.1	0.5	0.1	0.1
5	$A_{12} \sim A_{13}$	***1.6***	***2.1***	***2.1***	***1.9***	***2.1***	0.1	1.5	1.5	0.6	1.1	2.2
6	$A_{14} \sim A_{17}$	***1.0***	***1.0***	***1.0***	***0.9***	***1.0***	0.8	1.0	0.9	0.5	0.9	1.1
7	$A_{18} \sim A_{21}$	***0.4***	***0.5***	***0.5***	***0.5***	***0.5***	1.8	0.6	0.6	1.5	0.4	1.0
8	$A_{22} \sim A_{25}$	***3.4***	***3.4***	***3.4***	***3.4***	***3.4***	3.0	3.4	3.4	3.0	3.4	3.0
Weight (lb)		485.77 [521.04][a]	484.85 [504.72][a]	484.85 [514.20][a]	485.05 [514.21][a]	484.85 [504.28][a]	546.01	491.72	486.29	543.95	493.80	537.23
Number of structural analyses		13,736 [13,445][b]	14,163 [4,414][b]	13,523 [2,160][b]	17,159 [5,226][b]	18,734 [6,850][b]	600	-	40,000	-	-	-

[a] The HS optimal results obtained after 600 structural analyses (the result of Rajeev and Krishnamoorthy [21]).

[b] Number of analyses for the HS required to obtain a weight of 486.29 lb (the result of Wu and Chow, [27]).

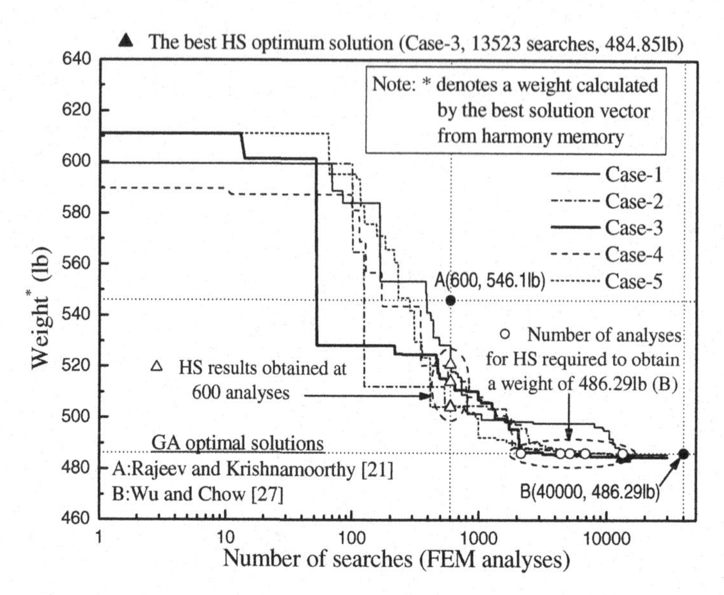

Fig. 16 Convergence History of Minimum Weight for 25-Bar Space Truss

weight) were obtained for all five HS cases (see Table 18). All of the HS results were better than the values obtained in the previous investigations.

Figure 16 showed a comparison of the convergence capability of each HS case and the GA-based approaches. While the pure GA proposed by Rajeev and Krishnamoorthy [21] obtained a minimum weight of 546.01 lb after 600 structural analyses, the HS cases obtained minimum weights of 504.28 to 521.04 lb after the same number of analyses. The steady-state GA proposed by Wu and Chow [27] obtained a minimum weight of 486.29 lb after 40,000 analyses, while all HS cases except Case 1 obtained the same weight after 2,160 to 6,850 analyses. These results suggest that the HS-based method is a powerful search and discrete size optimization technique, when compared to pure and steady-state GA-based methods, in terms of both the obtained optimal value and the convergence capability.

(2) 72-bar Space Truss

The 72-bar space truss, shown in Figure 9, is one of the most popular classical optimization design problems, and has been used as a benchmark to verify the efficiency of various optimization methods. The majority of these studies have assumed that the cross-sectional areas (size variables) were continuous. However, Wu and Chow [27] optimized this space structure with discrete cross-sectional areas using the steady-state GA-based method. In this example, the material density and modulus of elasticity were 0.1 lb/in.^{3} and 10,000 ksi, respectively. The space truss was subjected to the following two loading conditions: Condition 1, in which P_X = 5.0 kips, P_Y = 5.0 kips, and P_Z = -5.0 kips on node 17; and Condition 2, in which P_X = 0.0 kips, P_Y = 0.0 kips, and P_Z = -5.0 kips on nodes 17, 18, 19, and 20.

Table 20 Available Discrete Cross-Sections

No.	Areas	No.	Areas	No.	Areas	No.	Areas
1	0.111	17	1.563	33	3.840	49	11.500
2	0.141	18	1.620	34	3.870	50	13.500
3	0.196	19	1.800	35	3.880	51	13.900
4	0.250	20	1.990	36	4.180	52	14.200
5	0.307	21	2.130	37	4.220	53	15.500
6	0.391	22	2.380	38	4.490	54	16.000
7	0.442	23	2.620	39	4.590	55	16.900
8	0.563	24	2.630	40	4.800	56	18.800
9	0.602	25	2.880	41	4.970	57	19.900
10	0.766	26	2.930	42	5.120	58	22.000
11	0.785	27	3.090	43	5.740	59	22.900
12	0.994	28	3.130	44	7.220	60	24.500
13	1.000	29	3.380	45	7.970	61	26.500
14	1.228	30	3.470	46	8.530	62	28.000
15	1.266	31	3.550	47	9.300	63	30.000
16	1.457	32	3.630	48	10.850	64	33.500

Note: cross-sectional areas are in in.^{2}.

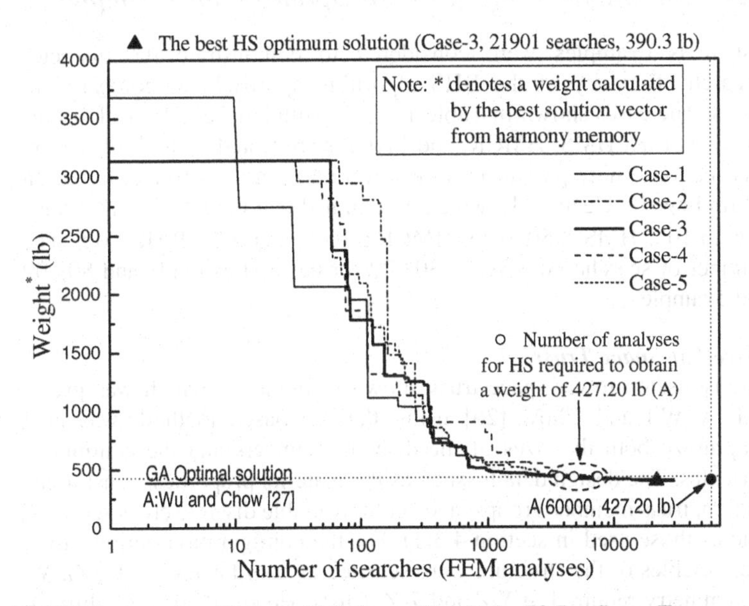

Fig. 17 Convergence History of Minimum Weight for 72-Bar Space Truss

The structure was required to be doubly symmetric about the *X*- and *Y*-axes. This condition divided the truss members into the following sixteen groups: (1) $A_1 \sim A_4$, (2) $A_5 \sim A_{12}$, (3) $A_{13} \sim A_{16}$, (4) $A_{17} \sim A_{18}$, (5) $A_{19} \sim A_{22}$, (6) $A_{23} \sim A_{30}$, (7) $A_{31} \sim A_{34}$, (8) $A_{35} \sim A_{36}$, (9) $A_{37} \sim A_{40}$, (10) $A_{41} \sim A_{48}$, (11) $A_{49} \sim A_{52}$, (12) $A_{53} \sim A_{54}$, (13) $A_{55} \sim A_{58}$, (14) $A_{59} \sim A_{66}$, (15) $A_{67} \sim A_{70}$, and (16) $A_{71} \sim A_{72}$. The members were subjected to stress limitations of ± 25 ksi, and the maximum displacement of the uppermost nodes was not allowed to exceed ± 0.25 in. for each node, in all directions. In this example, the available discrete values for the cross-sectional areas were chosen from the sixty-four discrete values listed in Table 20.

Table 21 gives each HS optimal result for the 72-bar space truss, along with several continuous optimal results and the results reported by Wu and Chow [27] using the GA-based discrete optimization method. The best minimum weight of 390.30 lb was obtained using the Case 3 parameters after 21,901 searches (FEM structural analyses). The results for each HS case shown in the table were better than the previous discrete design results reported by Wu and Chow [27]. Figure 17 showed a comparison of the convergence capability of each HS case and the steady-state GA-based method (Wu and Chow). Wu and Chow obtained a minimum weight of 427.2 lb after 60,000 structural analyses using a four-point crossover operator, while the proposed HS approach obtained the same weight after 3,711 to 7,462 analyses. The HS approach therefore outperformed the steady-state GA-based method, in terms of both the obtained optimal value and the convergence capability.

4.4 Discrete-Continuous Configuration Optimization Examples

Standard test truss examples were considered to demonstrate the discrete-continuous search efficiency of the BHS algorithm approach, as compared to current methods. The cases shown in Table 18, each with a different set of HS algorithm parameters (*i.e.*, HMS, HMCR, and PAR), were tested with all of the examples in this section. These parameter values were arbitrarily selected, based on the empirical findings by Geem [34], which determined that the HS algorithm performed well with $10 \leq \text{HMS} \leq 50$, $0.7 \leq \text{HMCR} \leq 0.95$, and $0.2 \leq \text{PAR} \leq 0.5$. The maximum number of searches was set to 30,000 for the first example and 80,000 for the second example.

(1) Twenty Five-bar Space Truss

The 25-bar transmission tower space truss shown in Figure 8, which was previously studied by Wu and Chow [26] using the GA-based method, was also analyzed to optimize both the sizes of the discrete members and the continuous geometric variables. The design details, such as the material properties, constraints, loading condition, truss member groups, and set of available discrete cross sections, were the same as those used in section 4.3(1). For the configuration optimization, the geometric variables of the structure were selected as coordinates X_4, Y_4, Z_4, X_8, and Y_8, with symmetry required in *X-Z* and *Y-Z* planes. Hence, there were thirteen independent design variables, including the eight sizing variables given in section 4.3(1) and five geometric variables. The side constraints for the geometric variables, *i.e.*, the lower and upper bounds on the nodal coordinates, were $20 \leq X_4 \leq 60$, $40 \leq Y_4 \leq 80$, $90 \leq Z_4 \leq 130$, $40 \leq X_8 \leq 80$, and $100 \leq Y_8 \leq 140$ (in.).

The HS-based discrete-continuous configuration optimization method was applied to the 25-bar space truss using each set of parameters shown in Table 18. The algorithm found the best solution vector (*i.e.*, the values of the eight sizing variables and five geometric variables) with each set of parameters within 30,000 searches. Table 22 gives the best solution and the corresponding minimum structural weight for

Table 21 Optimal Results for 72-bar space truss

	Design variables A_i (in.^{2})	HS results					Wu and Chow [27]				Xicheng & Guixu	Erbatur *et al.*
		Case-1	Case-2	Case-3	Case-4	Case-5	1X[b]	2X[b]	3X[b]	4X[b]	[68][c]	[31][c]
1	$A_1 \sim A_4$	1.800	1.990	1.990	1.990	1.620	1.563	1.990	1.990	1.563	1.905	1.910
2	$A_5 \sim A_{12}$	0.602	0.602	0.442	0.602	0.602	0.307	0.602	0.563	0.766	0.518	0.525
3	$A_{13} \sim A_{16}$	0.111	0.111	0.111	0.111	0.111	0.111	0.141	0.111	0.141	0.100	0.122
4	$A_{17} \sim A_{18}$	0.111	0.111	0.111	0.111	0.111	0.111	0.111	0.141	0.111	0.100	0.103
5	$A_{19} \sim A_{22}$	1.457	1.228	1.266	1.457	1.457	2.130	0.994	1.457	1.800	1.286	1.310
6	$A_{23} \sim A_{30}$	0.563	0.563	0.563	0.391	0.391	0.602	0.602	0.602	0.602	0.516	0.498
7	$A_{31} \sim A_{34}$	0.111	0.111	0.111	0.141	0.111	0.111	0.111	0.111	0.141	0.100	0.110
8	$A_{35} \sim A_{36}$	0.111	0.111	0.111	0.111	0.111	0.111	0.307	0.111	0.307	0.100	0.103
9	$A_{37} \sim A_{40}$	0.442	0.442	0.391	0.391	0.563	0.766	0.307	0.442	0.391	0.509	0.535
10	$A_{41} \sim A_{48}$	0.442	0.442	0.602	0.602	0.563	1.000	0.602	0.766	0.391	0.522	0.535
11	$A_{49} \sim A_{52}$	0.111	0.111	0.111	0.111	0.111	0.111	0.111	0.111	0.141	0.100	0.103
12	$A_{53} \sim A_{54}$	0.141	0.111	0.111	0.111	0.111	0.111	0.563	0.141	0.111	0.100	0.111
13	$A_{55} \sim A_{58}$	0.196	0.196	0.196	0.196	0.196	0.785	0.250	0.196	0.196	0.157	0.161
14	$A_{59} \sim A_{66}$	0.563	0.563	0.563	0.602	0.602	0.602	0.766	0.442	0.602	0.537	0.544
15	$A_{67} \sim A_{70}$	0.250	0.391	0.391	0.391	0.391	0.602	0.307	0.250	0.307	0.411	0.379
16	$A_{71} \sim A_{72}$	1.000	0.563	0.563	0.563	0.785	0.602	0.391	1.000	0.766	0.571	0.521
	Weight (lb)	400.63	390.62	390.30	399.23	396.38	471.98	439.77	428.00	427.20	380.84	383.12
	Number of structural analyses	25,717 [7,242][a]	26,812 [7,462][a]	21,901 [3,711][a]	13,866 [4,819][a]	22,894 [3,677][a]	60,000	60,000	60,000	60,000	-	-

[a] Number of analyses for the HS required to obtain a weight of 427.2 lb (the best result of Wu and Chow, [27]).

[b] Crossover operators used by Wu and Chow, [27].

[c] The optimal results of continuous size optimization.

each case, and also provides a comparison between the optimal design result reported by Wu and Chow [26] and the present work. The best minimum weight of 123.77 lb was obtained using the Case 5 parameters after 8,902 searches (structural analyses), and this minimum weight converged remarkably after only 2,000 searches. The results from each HS case were better than the previous design result reported by Wu and Chow, and the HS best result using the Case 5 parameters produced a weight saving of 10%, as compared to the GA-based method proposed by Wu and Chow.

The configuration optimization achieved an amazing optimal weight saving of 70%, as compared to the pure HS size optimization, which obtained a best minimum weight of only 484.85 lb, as shown in Table 19.

Table 22 Optimal Results of 25-Bar Space Truss

Design variables		HS results					Wu & Chow
A_i (in.^{2}) & R_i (in.)		Case-1	Case-2	Case-3	Case-4	Case-5	[26]
1	A_1	0.1	0.1	0.1	0.2	0.2	0.1
2	$A_2 \sim A_5$	0.2	0.1	0.2	0.2	0.1	0.2
3	$A_6 \sim A_9$	0.9	1.0	0.9	1.0	0.9	1.1
4	$A_{10} \sim A_{11}$	0.1	0.1	0.1	0.1	0.1	0.2
5	$A_{12} \sim A_{13}$	0.1	0.1	0.1	0.2	0.1	0.3
6	$A_{14} \sim A_{17}$	0.1	0.1	0.2	0.1	0.1	0.1
7	$A_{18} \sim A_{21}$	0.1	0.4	0.2	0.1	0.2	0.2
8	$A_{22} \sim A_{25}$	1.2	0.7	0.8	1.0	1.0	0.9
1	X_4	31.64	28.54	29.51	27.94	31.88	41.07
2	Y_4	66.30	55.18	56.76	55.21	53.57	53.47
3	Z_4	102.22	127.80	130.0	123.70	126.35	124.60
4	X_8	40.00	43.02	41.74	43.63	40.43	50.80
5	Y_8	125.74	136.66	133.62	130.83	130.64	131.48
Weight (lb)		129.34 [138.10][a] [130.40][b] [129.53][c] [129.36][d]	123.81 [152.10][a] [140.63][b] [134.29][c] [124.92][d]	126.07 [154.05][a] [141.65][b] [131.71][c] [131.03][d]	126.74 [168.09][a] [146.68][b] [133.87][c] [128.16][d]	123.77 [137.79][a] [124.28][b] [123.86][c] [123.80][d]	136.20
Number of structural analyses		29,290	9,646	23,100	19,833	8,902	-

[a] The structural weights obtained after 1,000 analyses.
[b] The structural weights obtained after 2,000 analyses.
[c] The structural weights obtained after 3,000 analyses.
[d] The structural weights obtained after 8,000 analyses.

(2) Forty Seven-bar Planar Power Line Tower

The 47-bar planar power line tower design, shown in Figure 18, was the last example used to demonstrate the practical capability of the HS algorithm-based structural optimization method. This tower was previously analyzed by Felix [61] and Hansen and Vanderplaats [65] to obtain optimal continuous size and geometric variables (*i.e.*, a continuous configuration optimization). In this problem, the structure had forty-seven members and twenty-two nodes, and was symmetric about the Y-axis. All members were made of steel, and the material density and modulus of elasticity were 0.3 lb/in.^{3} and 30,000 ksi, respectively.

This tower was designed for three separate load conditions: (1) 6.0 kips acting in the positive X-direction and 14.0 kips acting in the negative Y-direction at nodes 17 and 22, (2) 6.0 kips acting in the positive X-direction and 14.0 kips acting in the negative Y-direction at node 17, and (3) 6.0 kips acting in the positive X-direction and 14.0 kips acting in the negative Y-direction at node 22. The first condition represented the load imposed by two power lines attached to the tower at an angle. The second and third conditions represented cases that occur when one of the two lines snaps.

The structure was subjected to both stress and buckling constraints. The stress constraints were 15.0 ksi in compression and 20.0 ksi in tension. The Euler buckling compressive stress limit for each member i was used for the buckling constraints. This was computed as

$$\sigma_i^{cr} = \frac{-KEA_i}{L_i^2} \quad (i = 1,\ldots,47) \tag{11}$$

where K is a constant determined from the cross-sectional geometry, E is the modulus of elasticity of the material, and L_i is the member length. In this study, the buckling constant was $K = 3.96$.

The cross-sectional areas of the members were categorized into twenty-seven groups, as follows: (1) $A_1 = A_3$, (2) $A_2 = A_4$, (3) $A_5 = A_6$, (4) A_7, (5) $A_8 = A_9$, (6) A_{10}, (7) $A_{11} = A_{12}$, (8) $A_{13} = A_{14}$, (9) $A_{15} = A_{16}$, (10) $A_{17} = A_{18}$, (11) $A_{19} = A_{20}$, (12) $A_{21} = A_{22}$, (13) $A_{23} = A_{24}$, (14) $A_{25} = A_{26}$, (15) A_{27}, (16) A_{28}, (17) $A_{29} = A_{30}$, (18) $A_{31} = A_{32}$, (19) A_{33}, (20) $A_{34} = A_{35}$, (21) $A_{36} = A_{37}$, (22) A_{38}, (23) $A_{39} = A_{40}$, (24) $A_{41} = A_{42}$, (25) A_{43} , (26) $A_{44} = A_{45}$, and (27) $A_{46} = A_{47}$. The independent geometric variables were X_2, X_4, Y_4, X_6, Y_6, X_8, Y_8, X_{10}, Y_{10}, X_{12}, Y_{12}, X_{14}, Y_{14}, X_{20}, Y_{20}, X_{21}, and Y_{21}. The geometric variables were linked to maintain symmetry about the Y-axis. Nodes 1 and 2 were required to remain at $Y = 0.0$, and the coordinates of nodes 15, 16, 17, and 22 were not changed. There were forty-four independent design variables, including twenty-seven sizing variables and seventeen coordinate variables.

In this example, the cross-sectional areas were chosen from the sixty-four discrete values listed in Table 20, and a pure discrete sizing variable problem (with fixed geometry) was also optimized for comparison. Table 23 gives the optimal results obtained using each set of HS parameters for the discrete-continuous configuration optimization, along with the optimal results for the continuous configuration problem. The best pure discrete size result, which was obtained using the Case 3 parameters, is also listed in the table. After 73,257 to 76,937 searches (structural analyses), the best discrete-continuous solution vector and the corresponding objective function value were obtained for each HS case. The best minimum weight of 2,020.78 lb was obtained using the Case 1 parameters after 73,771 searches, and this minimum weight converged remarkably after 40,000 searches. The discrete-continuous configuration optimization produced a considerable weight saving of 16%, as compared to the pure discrete size optimization, which obtained a minimum weight of 2,396.8 lb.

Table 23 Optimal Results of 47-Bar Planar Power Line

Variables A_i (in.^{2}) & R_i (in.)	Pure Size Case-3*	HS results Case-1	Case-2	Case-3	Case-4	Case-5	Felix** [61]	Han. & Van.** [65]
1 $A_1 = A_3$	3.840	2.620	3.550	3.130	3.090	2.930	2.73	2.42
2 $A_2 = A_4$	3.380	2.630	3.090	3.090	2.880	2.630	2.47	2.35
3 $A_5 = A_6$	0.766	1.228	0.766	1.000	0.994	1.228	0.73	0.82
4 A_7	0.141	0.196	0.141	0.111	0.141	0.141	0.21	0.10
5 $A_8 = A_9$	0.785	1.000	1.000	0.994	1.228	0.994	0.94	0.86
6 A_{10}	1.990	1.620	1.228	1.228	1.620	1.800	1.08	1.15
7 $A_{11} = A_{12}$	2.130	1.800	1.990	2.130	2.380	2.380	1.69	1.77
8 $A_{13} = A_{14}$	1.228	0.785	1.000	0.785	0.602	0.602	0.69	0.67
9 $A_{15} = A_{16}$	1.563	1.000	1.228	1.228	1.228	0.994	1.06	0.86
10 $A_{17} = A_{18}$	2.130	1.563	1.800	1.990	1.620	1.620	1.41	1.24
11 $A_{19} = A_{20}$	0.111	0.391	0.602	0.785	0.563	0.602	0.26	0.33
12 $A_{21} = A_{22}$	0.111	0.766	0.994	0.994	1.457	1.228	0.81	1.22
13 $A_{23} = A_{24}$	1.800	1.228	1.457	1.457	1.228	1.228	1.06	0.93
14 $A_{25} = A_{26}$	1.800	1.228	1.457	1.457	1.228	1.228	1.05	0.86
15 A_{27}	1.457	1.228	1.228	1.000	1.457	1.457	0.82	0.69
16 A_{28}	0.442	0.196	0.250	0.111	0.141	0.196	0.30	0.15
17 $A_{29} = A_{30}$	3.630	2.930	2.880	2.880	3.130	3.130	2.77	2.46
18 $A_{31} = A_{32}$	1.457	0.994	1.228	0.994	0.994	0.766	0.66	0.90
19 A_{33}	0.391	0.111	0.111	0.141	0.111	0.111	0.21	0.10
20 $A_{34} = A_{35}$	3.090	3.470	2.880	3.130	3.380	3.550	2.90	2.74
21 $A_{36} = A_{37}$	1.457	1.000	1.000	1.228	1.000	1.000	0.27	0.92
22 A_{38}	0.196	0.111	0.111	0.307	0.111	0.111	1.41	0.10
23 $A_{39} = A_{40}$	3.840	3.380	3.130	3.380	3.470	3.380	3.43	2.94
24 $A_{41} = A_{42}$	1.563	1.228	1.266	1.000	1.228	1.000	0.99	1.13
25 A_{43}	0.196	0.111	0.111	0.111	0.250	0.111	0.17	0.10
26 $A_{44} = A_{45}$	4.590	3.380	3.470	3.630	3.470	3.470	3.65	3.12
27 $A_{46} = A_{47}$	1.457	0.994	1.563	1.266	0.994	1.266	1.01	1.10
1 $-X_1=X_2$	60.0*	98.9	89.4	85.9	97.7	91.6	90.0	107.1
2 $-X_3= X_4$	60.0*	80.9	83.1	80.9	80.9	80.9	90.0	91.2
3 $Y_3=Y_4$	120.0*	114.8	111.7	115.4	114.1	122.9	123.4	122.8
4 $-X_5=X_6$	60.0*	62.8	74.4	61.6	60.1	61.6	83.4	74.2
5 $Y_5=Y_6$	240.0*	236.9	234.9	233.9	225.2	238.4	244.5	241.4
6 $-X_7=X_8$	60.0*	51.3	59.5	55.4	49.2	47.6	70.5	65.5
7 $Y_7=Y_8$	360.0*	315.9	339.3	319.4	323.1	327.9	355.1	324.6
8 $-X_9=X_{10}$	30.0*	47.9	40.7	46.9	44.4	41.1	60.0	57.1
9 $Y_9=Y_{10}$	420.0*	387.4	429.6	409.9	392.1	394.1	425.0	400.4
10 $-X_{11}=X_{12}$	30.0*	50.3	35.2	35.3	38.1	42.7	58.2	49.3
11 $Y_{11}=Y_{12}$	480.0*	477.3	455.3	471.8	477.3	476.6	478.0	472.3
12 $-X_{13}=X_{14}$	30.0*	41.4	34.4	36.6	40.1	42.4	59.6	47.4
13 $Y_{13}=Y_{14}$	540.0*	521.4	505.7	504.9	519.2	504.7	519.5	507.5
14 $-X_{18}=X_{21}$	90.0*	92.5	83.9	84.8	91.3	84.7	96.9	83.3
15 $Y_{18}=Y_{21}$	600.0*	615.3	609.2	606.8	620.9	615.5	633.7	636.0
16 $-X_{19}=X_{20}$	30.0*	14.3	18.4	17.1	6.9	3.2	15.0	3.9
17 $Y_{19}=Y_{20}$	600.0*	596.5	586.4	582.0	580.7	569.6	607.6	586.5
Weight (lb)	2,396.8 [2,471.1][a] [2,434.3][b] [2,407.7][c]	2,020.78 [2,428.6][a] [2,198.1][b] [2,066.7][c]	2,116.14 [2,608.2][a] [2,339.8][b] [2,195.2][c]	2,091.21 [2,580.5][a] [2,361.1][b] [2,189.5][c]	2,096.35 [2,735.43][a] [2,421.92][b] [2,225.39][c]	2,056.77 [2,468.82][a] [2,269.06][b] [2,165.42][c]	1,904.0	1,850.4
Number of analyses	45,557	73,771	76,937	74,721	76,828	73,257	-	-

* Coordinate is stationary. ** The results of continuous configuration optimizations.

[a] Structural weights obtained after 10,000 analyses.

[b] Structural weights obtained after 20,000 analyses.

[c] Structural weights obtained after 40,000 analyses.

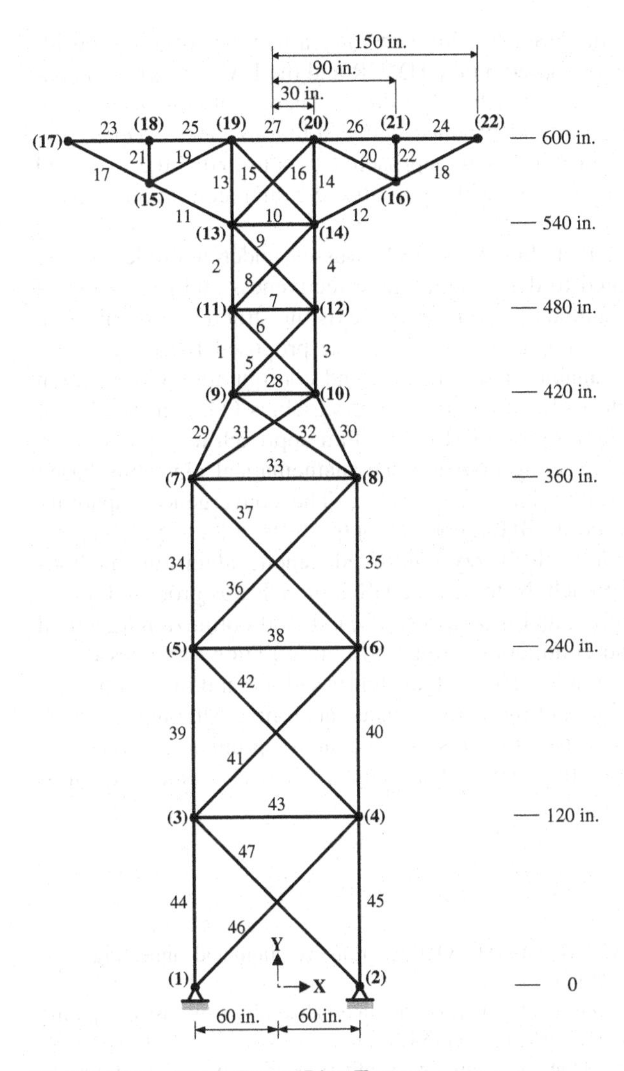

Fig. 18 47-Bar Planar Power Line Tower

5 Conclusion Remarks

The HS meta-heuristic algorithm was conceptualized using the musical process of searching for a perfect state of harmony. Compared to gradient-based mathematical optimization algorithms, the HS algorithm imposes fewer mathematical requirements to solve optimization problems and does not require initial starting values for the decision variables.

The HS algorithm uses a stochastic random search [69, 70] based on the harmony memory considering rate (HMCR) and pitch adjusting rate (PAR), which effectively guide a global search, and calculus-based derivative information is unnecessary. Furthermore, the HS algorithm generates a new vector after considering all of the existing vectors based on the HMCR and the PAR, rather than considering only two (parents) as in genetic algorithms. These features increase the flexibility of the HS algorithm and produce better solutions. This chapter presented the original HS algorithm-based approach for optimizing the size and configuration of structural systems with both discrete and continuous design variables.

Various truss examples, including large-scale trusses under multiple loading conditions, are also presented to demonstrate the effectiveness and robustness of the BHS algorithm-based methods, as compared to existing structural optimization techniques. The numerical examples revealed that the proposed BHS algorithm-based search strategy was capable of solving size and configuration optimization problems with both discrete and continuous design variables Optimal weights of structures obtained using the proposed BHS algorithm approach may yield better solutions than those obtained using conventional mathematical algorithm-based approaches or genetic algorithm-based approaches. The convergence capability results revealed that the proposed BHS approach outperformed the simple genetic algorithm-based method, while the fuzzy controlled genetic algorithm methods were better than the HS approach. Note that the HS approach was proposed on the basis of the pure HS algorithm, and is a powerful search and optimization method for solving the discrete and continuous sizing and configuration variables of the structures compared to the simple genetic algorithm-based method in term of both the obtained optimal solution and the convergence capability. Although the proposed approach is applied to truss structures, it is a general optimization procedure that can be easily used for other types of structures, such as frame structures, plates, and shells.

References

1. Kirkpatrick, S., Gelatt, C., Vecchi, M.: Optimization by simulated annealing. Science 220(4598), 671–680 (1983)
2. Metropolis, et al.: Metropolis et al Equations of state calculations by fast computing machines. J. Chem. Phys. 21, 1087–1092 (1953)
3. Glover, F.: Heuristic for integer programming using surrogate constraints. Decision Science 8(1), 156–166 (1977)
4. Holland, J.H.: Adaptation in natural and artificial systems. University of Michigan Press, Ann Arbor (1975)
5. Goldberg, D.E.: Genetic algorithm in search optimization and machine learning. Addision Wesley, Boston (1989)
6. Schwefel, H.P.: On the evolution of evolutionary computation. In: Zurada, J., Marks, R., Robinson, C. (eds.) Computational intelligence: imitating life, pp. 116–124. IEEE Press, New York (1994)
7. Fogel, D.B.: A comparison of evolutionary programming and genetic algorithms on selected constrained optimization problems. Simulation 64(6), 399–406 (1995)
8. Fogel, L.J., Owens, A.J., Walsh, M.J.: Artificial intelligence through simulated evolution. John Wiley, Chichester (1966)

9. Koza, J.R.: Genetic programming: A paradigm for genetically breeding populations of computer programs to solve problems. Rep. No. STAN-CS-90-1314, Stanford University, CA (1990)
10. Adeli, H., Cheng, N.T.: Integrated genetic algorithm for optimization of space structures. J. Aerospace Engineering, ASCE 6(4), 315–328 (1993)
11. Hajela, P.: Genetic search: An approach to the non-convex optimization problem. J. AIAA 28(7), 1205–1210 (1990)
12. Jenkins, W.M.: Towards structural optimization via the genetic algorithm. Comput. Struct. 40(5), 1321–1327 (1991)
13. Jenkins, W.M.: Structural optimization with the genetic algorithm. Struct. Engineer. 69(24), 418–422 (1991)
14. Jenkins, W.M.: Plane frame optimum design environment based on genetic algorithm. J. Struct. Engrg., ASCE 118(11), 3103–3112 (1992)
15. Jenkins, W.M.: A genetic algorithm for structural design optimization. In: Grierson, D.E., Hajela, P. (eds.) Emergent computing methods in engineering design: Application of genetic algorithms and neural network, pp. 30–53. Springer, Berlin (1996)
16. Grierson, D.E., Pak, W.H.: Optimal sizing, geometrical and topological design using a genetic algorithm. Struct. Optimization 6, 151–159 (1993)
17. Ohsaki, M.: Genetic algorithm for topology optimization of trusses. Comput. Struct. 57(2), 219–225 (1995)
18. Rajan, S.D.: Sizing, shape, and topology design optimization of trusses using genetic algorithms. J. Struct. Engrg. ASCE 121(10), 1480–1487 (1995)
19. Yang, J.P., Soh, C.K.: Structural optimization by genetic algorithms with tournament selection. J. Comp. in Civ. Engrg. ASCE 11(3), 195–200 (1997)
20. Galante, M.: Genetic algorithms as an approach to optimize real-world trusses. Int. J. Numer. Methods Engrg. 39, 361–382 (1996)
21. Rajeev, S., Krishnamoorthy, C.S.: Discrete optimization of structures using genetic algorithms. J. Structural Engineering, ASCE 118(5), 1233–1250 (1992)
22. Rajeev, S., Krishnamoorthy, C.S.: Genetic algorithm-based methodologies for design optimization of trusses. J. Structural Engineering, ASCE 123(3), 350–358 (1997)
23. Koumousis, V.K., Georgious, P.G.: Genetic algorithms in discrete optimization of steel truss roofs. J. Computing in Civil Engineering, ASCE 8(3), 309–325 (1994)
24. Hajela, P., Lee, E.: Genetic algorithms in truss topological optimization. Int. J. Solids and Structures 32(22), 3341–3357 (1995)
25. Adeli, H., Kumar, S.: Distributed genetic algorithm for structural optimization. J. Aerospace Engineering, ASCE 8(3), 156–163 (1995)
26. Wu, S.-j., Chow, P.-T.: Integrated discrete and configuration optimization of trusses using genetic algorithms. Computers and Structures 55(4), 695–702 (1995)
27. Wu, S.-j., Chow, P.-T.: Steady-state genetic algorithms for discrete optimization of trusses. Computers and Structures 56(6), 979–991 (1995)
28. Soh, C.K., Yang, J.: Fuzzy controlled genetic algorithm search for shape optimization. J. Computing in Civil Engineering, ASCE 10(2), 143–150 (1996)
29. Camp, C., Pezeshk, S., Cao, G.: Optimized design of two-dimensional structures using a genetic algorithm. J. Structural Engineering, ASCE 124(5), 551–559 (1998)
30. Shrestha, S.M., Ghaboussi, J.: Evolution of optimization structural shapes using genetic algorithm. J. Structural Engineering, ASCE 124(11), 1331–1338 (1998)
31. Erbatur, F., Hasancebi, O., Tutuncil, I., Kihc, H.: Optimal design of planar and structures with genetic algorithms. Computers and Structures 75, 209–224 (2000)

32. Sarma, K.C., Adeli, H.: Fuzzy genetic algorithm for optimization of steel structures. J. Structural Engineering, ASCE 126(5), 596–604 (2000)
33. Haftka, R.T., Le Riche, R., Harrison, P.: Genetic algorithms for the design of composite panels. In: Grierson, D.E., Hajela, P. (eds.) Emergent computing methods in engineering design: Application of genetic algorithms and neural network, pp. 10–29. Springer, Berlin (1996)
34. Geem, Z.W., Kim, J.-H., Loganathan, G.V.: A new heuristic optimization algorithm: harmony search. Simulation 76(2), 60–68 (2001)
35. Pezeshk, S., Camp, C.V., Chen, D.: Design of nonlinear framed structures using genetic optimization. J. Struct. Engrg., ASCE 126(3), 382–388 (2000)
36. Schmit Jr., L.A., Farshi, B.: Some approximation concepts for structural synthesis. AIAA J. 12(5), 692–699 (1974)
37. Schmit Jr., L.A., Miura, H.: Approximation concepts for efficient structural synthesis. NASA CR-2552, Washington, D. C (1976)
38. Venkayya, V.B.: Design of optimum structures. Computers and Structures 1(1-2), 265–309 (1971)
39. Gellatly, R.A., Berke, L.: Optimal structural design. AFFDL-TR-70-165, Air Force Flight Dynamics Lab., Wright-Patterson AFB, OH (1971)
40. Dobbs, M.W., Nelson, R.B.: Application of optimality criteria to automated structural design. AIAA J. 14(10), 1436–1443 (1976)
41. Rizzi, P.: Optimization of multiconstrained structures based on optimality criteria. In: AIAA/ASME/SAE 17th Structures, Structural Dynamics, and Materials Conference, King of Prussia, PA (1976)
42. Khan, M.R., Willmert, K.D., Thornton, W.A.: An optimality criterion method for large-scale structures. AIAA J. 17(7), 753–761 (1979)
43. John, K.V., Ramakrishnan, C.V., Sharma, K.G.: Minimum weight design of truss using improved move limit method of sequential linear programming. Computers and Structures 27(5), 583–591 (1987)
44. Sunar, M., Belegundu, A.D.: Trust region methods for structural optimization using exact second order sensitivity. Int. J. Numerical Methods in Engineering 32, 275–293 (1991)
45. Stander, N., Snyman, J.A., Coster, J.E.: On the robustness and efficiency of the S.A.M. algorithm for structural optimization. Int. J. Int. J. Numerical Methods in Engineering 38, 119–135 (1995)
46. Xu, S., Grandhi, R.V.: Effective two-point function approximation for design optimization. AIAA J. 36(12), 2269–2275 (1998)
47. Lamberti, L., Pappalettere, C.: Comparison of the numerical efficiency of different sequential linear programming based on algorithms for structural optimization problems. Computers and Structures 76, 713–728 (2000)
48. Lamberti, L., Pappalettere, C.: Move limits definition in structural optimization with sequential linear programming - PartII: Numerical examples. Computers and Structures 81, 215–238 (2003)
49. Khot, N.S., Berke, L.: Structural optimization using optimality criteria methods. In: Atrek, E., Gallagher, R.H., Ragsdell, K.M., Zienkiewicz, O.C. (eds.) New directions in optimum structural design. John Wiley & Sons, New York (1984)
50. Imai, K., Schmit Jr., L.A.: Configuration optimization of trusses. J. Structural Division. ASCE 107(ST5), 745–756 (1981)
51. Sheu, C.Y., Schmit Jr., L.A.: Minimum weight design of elastic redundant trusses under multiple static load conditions. AIAA J. 10(2), 155–162 (1972)

52. Templeman, A.B., Winterbottom, S.K.: Structural design by geometric programming. In: Second symposium on structural optimization, AGARD Conference, Preprint-123, Milan (1973)
53. Chao, N.H., Fenves, S.J., Westerberg, A.W.: Application of reduced quadratic programming technique to optimal structural design. In: Atrex, E., Gallagher, R.H., Radgsdell, K.M., Zienkiewicz, O.C. (eds.) New Directions in Optimum Structural Design. John Wiley, New York (1984)
54. Adeli, H., Kamal, O.: Efficient optimization of space trusses. Computers and Structures 24(3), 501–511 (1986)
55. Saka, M.P.: Optimum design of pin-jointed steel structures with practical applications. J. Structural Engineering, ASCE 116(10), 2599–2620 (1990)
56. Fadel, G.M., Clitalay, S.: Automatic evaluation of move-limits in structural optimization. Structural Optimization 6, 233–237 (1993)
57. Berke, L., Khot, N.S.: Use of optimality criteria methods for large-scale systems. AGARD Lecture Series No. 70 on Structural Optimization AGARD-LS-70, 1–29 (1974)
58. Xicheng, W., Guixu, M.: A parallel iterative algorithm for structural optimization. Computer Methods in Applied Mechanics and Engineering 96, 25–32 (1992)
59. Adeli, H., Park, H.-S.: Neurocomputing for design automation. CRC Press, Boca Raton (1998)
60. American Institute of Steel Construction (AISC), Manual of steel construction - allowable stress design. 9th edn., Chicago, Ill (1989)
61. Felix, J.E.: Shape optimization of trusses subjected to strength, displacement, and frequency constraints. Master's Thesis, Naval Postgraduate School (1981)
62. Yang, J.P.: Development of genetic algorithm-based approach for structural optimization. PhD Thesis, Nanyang Technol. Univ., Singapore (1996)
63. Yang, J.P., Soh, C.K.: Structural optimization by genetic algorithms with tournament selection. J. Comp. In: Civ. Engrg., ASCE 11(3), 195–200 (1997)
64. Vanderplaats, G.N., Moses, F.: Automated design of trusses for optimum geometry. J. Struct. Div., ASCE 98(3), 671–690 (1972)
65. Hansen, S.R., Vanderplaats, G.N.: Approximation method for configuration optimization of trusses. J. AIAA 28(1), 161–168 (1990)
66. Adeli, H., Park, H.-S.: Hybrid CPN-neural dynamics model for discrete optimization of steel structures. Microcomputer Civil Engineering 11(5), 355–366 (1996)
67. Park, H.-S., Sung, C.-W.: Optimization of steel structures using distributed simulated annealing algorithm on a cluster of personal computers. Comp. Struct. 80, 1305–1316 (2002)
68. Xicheng, W., Guixu, M.: A parallel iterative algorithm for structural optimization. Computer Methods in Applied Mechanics and Engineering 96, 25–32 (1992)
69. Geem, Z.W.: Novel derivative of harmony search algorithm for discrete design variables. Applied Mathematics and Computation 199(1), 223–230 (2008)
70. Geem, Z.W.: Music-inspired harmony search algorithm: theory and applications. Springer, Berlin (2009)

Optimum Design of Steel Frames via Harmony Search Algorithm

S.O. Degertekin[1]

Abstract. A harmony search algorithm is presented for optimum design of planar and space steel frames in this chapter. Harmony search (HS) is a meta-heuristic search method. It bases on the analogy between natural musical performance process and searching the solutions to optimization problems. The design algorithm aims to obtain minimum weight frames by selecting a standard set of steel sections. Strength constraints of AISC Load and Resistance Factor Design (LRFD) specification, displacement constraints and also size constraint for columns were imposed on frames. The effectiveness and robustness of harmony search algorithm, in comparison with genetic algorithm, simulated annealing and colony optimization based methods, were verified using three planar and two space steel frames. The comparisons showed that the harmony search algorithm yielded lighter designs for the presented examples.

1 Introduction

Computer-aided optimization has been used to obtain more economical designs since 1970s. Numerous algorithms have been developed for accomplishing the optimization problems in the last four decades. The early works on the topic mostly use mathematical programming techniques or optimality criteria with continuous design variables. These methods utilize gradient of functions to search the design space, but they are prone to converge locally optimum solutions. Furthermore, they are largely suitable for optimization problems with continuous design variables and they are not good enough for problems with discrete design variables. However, the availability of standard steel sections and their limitations for construction and manufacturing reasons necessitate that design variables selections be made from standard steel section lists recommended by design codes.

A number of articles were reported for the optimum design of structural systems [1-5]. A few articles deal with the optimum design of structures subjected to actual design constraints of code specifications [6-9]. Calculus-based and optimality criteria methods with continuous design variables were used in all these articles.

Today's competitive world has forced the engineers to realize more economical designs and designers to search/develop more effective optimization techniques. As a result, heuristic search methods emerged in the first half of 1990s. Many

[1] S.O. Degertekin
Department of Civil Engineering, Dicle University, 21280, Diyarbakir, Turkey
Email: sozgur@dicle.edu.tr

Z.W. Geem (Ed.): Harmony Search Algo. for Structural Design Optimization, SCI 239, pp. 51–78.
springerlink.com

heuristic search algorithms have been applied to various optimum design problems since then. Genetic algorithms (GAs), simulated annealing (SA) and ant colony optimization (ACO) that appeared as optimization tools are quite effective in obtaining the optimum solution of discrete optimization problems. One of the applications of heuristic search methods is optimum design of steel frames.

Broadly speaking, all heuristic search algorithms are inspired from natural phenomenon. The name of each heuristic method is indicative of its underlying principle. GAs are based on evolution theory of Darwin's. They were proposed by Holland [10]. The main principle of GAs is the survival of robust ones and the elimination of the others in a population. GAs are able to deal with discrete optimum design problems and do not need derivatives of functions, unlike classical optimization. However, the procedure for the genetic algorithm is time consuming and the optimum solutions may not be global ones, but they are feasible both mathematically and practically. GAs have been employed to solve many structural optimization problems since 1990s. They were used for the optimum design of planar/space trusses and frames [11-20]. GAs were also used to obtain optimum design of semi-rigid steel frames under the actual constraints of design codes [21-25].

SA is an accepted local search optimization method. Local search is an emerging paradigm for combinatorial search which has recently been shown to be very effective for a large number of combinatorial problems. It is based on the idea of navigating the search space by iteratively stepping from one solution to one of its neighbours, which are obtained by applying a simple local change to it. The SA algorithm is inspired by the analogy between the annealing of solids and searching the solutions to optimization problems. Annealing is a thermal process applied to solids by heating up them to a maximum temperature value at which all molecules of the solid crystal randomly arrange themselves in the liquid phase. The temperature of the molten crystal falls slowly later on. The crystalline structure becomes very tidy if the maximum temperature is quite high and the cooling is performed slowly enough. In this case, all molecules arrange themselves in the lowest energy (ground state). An analogy between the annealing and the optimization can be established in the following way: the energy of the solid denotes the objective (cost) function while the different states (configurations) during the cooling represent the different solutions (designs) throughout the optimization process. SA was developed by Metropolis et al. [26] and proposed by Kirkpatrick et al. [27] for optimization problems. Detailed explanation and different applications about SA can be found in the book by van Laarhoven and Aarts [28]. SA was also applied to the optimum design of steel frames under the actual design constraints and loads of code specifications [29-34].

ACO is an application of ant behaviour to the computational algorithms and is able to solve discrete optimum structural problems. It also has additional artificial characteristics such as memory, visibility and discrete time. ACO was originally put forward by Dorigo et al. [35] for optimization problems. The applications of ACO to the structural optimization were about the optimal design of planar/space trusses and frames [36-38].

A new meta-heuristic search algorithm called harmony search has been developed recently. Harmony search (HS) bases on the analogy between the performance process of natural music and searching for solutions to optimization problems. HS was developed by Geem et al. [39] for solving combinatorial optimization problems. HS can be easily programmed and adopted for engineering problems. Although HS is a relatively new heuristic algorithm, it has been applied to a diverse range of engineering problems. These are: river flood model [40], a conceptual rainfall-runoff model considering seasonal variation [41], vehicle routing [42], optimal design of water distribution networks [43], optimal scheduling of multiple dam system [44], ecological optimization [45], bandwidth-delay-constrained least-cost multicast routing [46] and minimization for slope stability analysis [47]. As regards the using of HS in the optimization of structural systems, the following articles can be considered: Lee and Geem [48] used HS algorithm for planar/space truss optimization with continuous design variables, Lee et al. [49] optimized the truss structures with discrete design variables, Lee and Geem [50] applied the HS algorithm for continuous engineering optimization, Saka [51,52] presented the HS algorithm for optimum geometry design of geodesic domes and steel frames. Degertekin [53,54] reported optimized designs for planar/space steel frame structures under the actual design constraints and loads of code specifications using HS algorithm.

The main differences between HS and GA are summarized as: (i) HS generates a new design considering all existing designs, while GA generates a new design from a couple of chosen parents by exchanging the artificial genes; (ii) HS takes into account each design variable independently. On the other hand, GA considers design variables depending upon building block theory [55]. (iii) HS does not code the parameters, whereas GA codes the parameters. That is, HS uses real value scheme, while GA uses binary scheme (0 and 1). The main differences between HS and SA are also summarized as: (i) HS obtains a new design considering all existing designs as mentioned above, while SA generates a new design considering few neighbour designs of current design. (ii) HS preserves better designs in its memory whereas SA does not have memory facility. HS has following advantages in comparison with ACO: (i) ACO develops new designs considering the collective information obtained from the pheromone trails of ants, while HS develops the new designs considering the former designs stored in its memory, similar to ACO, but it also takes into account all the design variable databases with a predetermined probability. This facility provides a chance to improve the design by the values not stored in HS memory. (ii) Local search process is applied to each other design with a predetermined probability in the HS, whereas ACO uses local search for only some elite designs. (iii) HS updates its memory after each design is generated. Therefore, the next design is obtained using updated harmony memory. On the other hand, ant colony is updated after as many designs as the number of ants in the colony are performed. These differences provide a more effective and powerful approach for HS than GA, SA and ACO.

In this chapter, the HS algorithm is applied to optimum design of planar/space steel frames under the actual design constraints of code specifications AISC-LRFD [56]. The design algorithm aims to obtain minimum weight frames by selecting a standard set of steel sections such as American Institute of Steel Construction (AISC) wide-flange (W) shapes. Strength constraints of AISC Load and Resistance Factor Design (LRFD) specification, displacement constraints and also size constraint for columns were imposed on frames. The versatility and robustness of HS algorithm, when compared to GA, SA and ACO based algorithms, were verified using 15-member, 30-member, 168-member planar frames and 8-member, 84-member space frames taken from current literature.

The formulations of the optimum design problem are presented in section 2. The basis of the HS algorithm, which addresses steel frames, is clarified in Section 3. Optimum design using HS algorithm is explained in section 4. Section 5 proposes the HS algorithm for optimum design of steel frames. Several examples of planar and space steel frames are given in section 6. Finally, section 7 draws some concluding remarks on this research.

2 The Formulations of the Optimum Design Problem

Formulation of an optimum design problem involves transcribing a verbal description of the problem into a well-defined mathematical statement [57]. A set of variables to describe the design, called design variables, are given in the formulation. All designs have to satisfy a given set of constraints which include limitations on material sizes and response of the system. If a design satisfies all constraints, it is accepted as a feasible design. A criterion is needed to decide whether or not one design is better than another. This criterion is called the objective function.

General flowchart diagram for optimum design could be sketched as shown in Figure 1 [57].

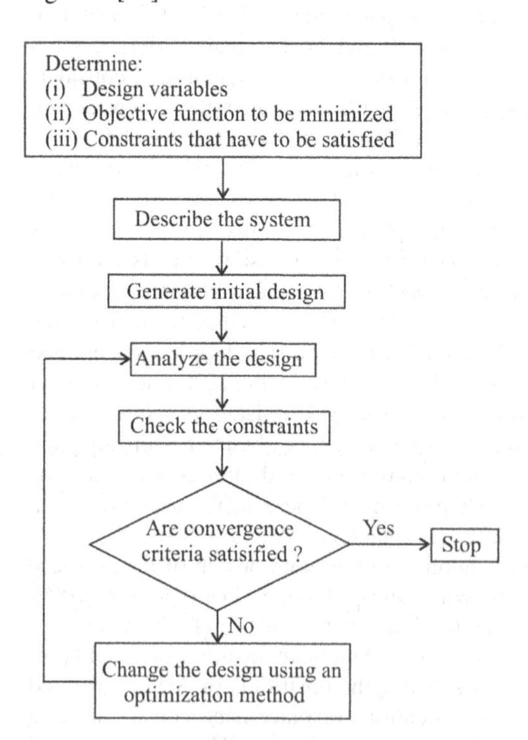

Fig. 1 General flowchart diagram for optimum design

The minimum weight could be considered as the objective function, the standard steel sections are treated as design variables and the constraints are taken from the design codes. Therefore; the discrete optimum design problem of steel frames can be stated as follows

$$\text{Minimize } W(x) = \sum_{k=1}^{ng} A_k \sum_{i=1}^{mk} \rho_i L_i \tag{1}$$

subjected to the strength constraints of AISC-LRFD [56] and displacement constraints. In Eqn. (1), *mk* is the total numbers of members in group *k*, ρ_i and L_i are density and length of member *i*, A_k is cross-sectional area of member group *k*, and *ng* is total numbers of groups in the frame.

The unconstrained objective function $\varphi(x)$ is then written for AISC-LRFD [56] code as

$$\varphi(x) = W(x)[1 + \kappa C]^{\varepsilon} \tag{2}$$

where *C*=constraint violation function, κ =penalty constant, ε=penalty function exponent. The constraint violation function

$$C = \sum_{i=1}^{N_j} C_i^t + \sum_{i=1}^{N_s} C_i^d + \sum_{i=1}^{N_{cl}} C_i^s + \sum_{i=1}^{N_c} C_i^I \tag{3}$$

where C_i^d and C_i^I = constraint violations for displacement and the interaction formulas of the LRFD specification; N_j= number of joints in the top storey. N_s and N_c= number of storeys and number of beam columns, respectively. N_{cl} = the total number of columns in the frame except the ones at the bottom floor. The penalty may be expressed as

$$C_i = \begin{cases} 0 & if \quad \lambda_i \le 0 \\ \lambda_i & if \quad \lambda_i > 0 \end{cases} \tag{4}$$

The displacement constraints are

$$\lambda_i^t = \frac{|d_t|}{|d_t^u|} - 1.0 \tag{5}$$

$$\lambda_i^d = \frac{|d_i|}{|d_i^u|} - 1.0 \tag{6}$$

where d_t : maximum displacement in the top storey, d_t^u : allowable top storey displacement, d_i : interstorey displacement in storey *i*, d_i^u : allowable interstorey displacement (storey height/300).

The size constraint employed for constructional reasons is given as follows

$$\lambda_i^s = \frac{d_{un}}{d_{bn}} - 1.0 \tag{7}$$

where d_{un} and d_{bn} are depths of steel sections selected for upper and lower floor columns.

The strength constraints taken from AISC-LRFD [56] are expressed in the following equations.

For members subject to bending moment and axial force

$$\lambda_i^I = \left(\frac{P_u}{\phi P_n}\right) + \frac{8}{9}\left(\frac{M_{ux}}{\phi_b M_{nx}} + \frac{M_{uy}}{\phi_b M_{ny}}\right) - 1.0 \quad \text{for} \quad \frac{P_u}{\phi P_n} \geq 0.2 \tag{8}$$

$$\lambda_i^I = \left(\frac{P_u}{2\phi P_n}\right) + \left(\frac{M_{ux}}{\phi_b M_{nx}} + \frac{M_{uy}}{\phi_b M_{ny}}\right) - 1.0 \quad \text{for} \quad \frac{P_u}{\phi P_n} < 0.2 \tag{9}$$

where P_u = required axial strength (compression or tension), P_{n} = nominal axial strength (compression or tension), M_{ux} = required flexural strengths about the major axis, M_{uy} = required flexural strengths about the minor axis, M_{nx} = nominal flexural strength about the major axis, M_{ny} = nominal flexural strength about the minor axis (for two-dimensional frames, M_{uy} = 0), $\phi = \phi_c$ = resistance factor for compression (equal to 0.85), $\phi = \phi_t$ = resistance factor for tension (equal to 0.90), ϕ_b =flexural resistance factor (equal to 0.90).

The AISC-LRFD [56] design strength of columns is $\phi_c P_n$, where $P_n = A_g F_{cr}$ with F_{cr} given by

$$F_{cr} = \begin{cases} 0.658^{\lambda_c^2} F_y & 0 \leq \lambda_c \leq 1.5 \\ \dfrac{0.877}{\lambda_c^2} F_y & \lambda > 1.5 \end{cases} \tag{10}$$

in which

$$\lambda_c = \frac{KL}{r\pi}\sqrt{\frac{F_y}{E}} \tag{11}$$

where A_g = cross-sectional area of member, F_{cr}= critical compressive stress, λ_c= column slenderness parameter, F_y= yield stress of steel, K=effective-length factor, L= member length, r= governing radius of gyration, E= modulus of elasticity. The effective length factor K, for unbraced frames is calculated from the following approximate equation taken from Dumonteil [58].

$$K=\sqrt{\frac{1.6G_A G_B+4.0(G_A+G_B)+7.50}{G_A+G_B+7.50}} \tag{12}$$

where subscripts A and B denote the two ends of the column under consideration. The restraint factor G is stated as

$$G=\frac{\sum(I_c/L_c)}{\sum(I_g/L_g)} \tag{13}$$

where I_c is the moment of inertia and L_c is the unsupported length of a column section; I_g is the moment of inertia and L_g is unsupported length of a girder. Σ indicates a summation for all members rigidly connected to that joint (A or B) and lying in the plane of buckling of the column under consideration.

The specification provides for three methods of analysis by which the required flexural capacity M_u may be evaluated [59]: (i) M_u may be determined from a plastic analysis, (ii) M_u may be determined from a geometrically nonlinear analysis using the factored loads, (iii) M_u may be determined applying the moment magnification factors to consider the second order effects which is also a choice in lieu of the geometrically nonlinear analysis according to the AISC-LRFD [56].

Design strength of beams is $\phi_b M_n$. As long as $\lambda \leq \lambda_p$, the M_n is equal to M_p and the shape is compact. The plastic moment M_p is calculated from the equation

$$M_p = ZF_y \tag{14}$$

where Z= the plastic section modulus, λ_p= slenderness parameter to attain M_p. Details of the formulations are given in the AISC LRFD [56]. Broad information can be also found in the books by Gaylord et al. [59] and Galambos et al. [60].

3 Harmony Search

Harmony is defined as an attractive sound made by two or more notes being played at the same time. Do, Re, Mi, Fa, Sol, La, and Si are called notes which represent a single sound. The HS algorithm imitates musical improvisation process where the musicians try to find a better harmony. All musicians always desire to attain the best harmony, which could be accomplished by numerous practices. The pitches of the instruments are changed after the each practice.

Figure 2 illustrates the analogy between music improvisation and steel design. As explained by Lee and Geem [50], harmony memory (HM) is the most important part of HS. Jazz improvisation is the best example for clarifying the harmony memory. Many jazz trios consist of a guitarist, double bassist and pianist. Each musician in the trio has different pitches: guitarist [Fa, Mi, La, Sol, Do]; double bassist [Mi, Do, La, Si, Re]; pianist [Si, Re, Mi, La, Do]. Let guitarist randomly play Sol out of its pitches [Fa, Mi, La, *Sol*, Do], double bassist Si out of [Mi, Do, La, *Si*, Re] and pianist Re [Si, *Re*, Mi, La, Do]. Therefore, the new harmony [Sol, Si, Re] becomes another harmony (musically G-chord). If the new harmony is

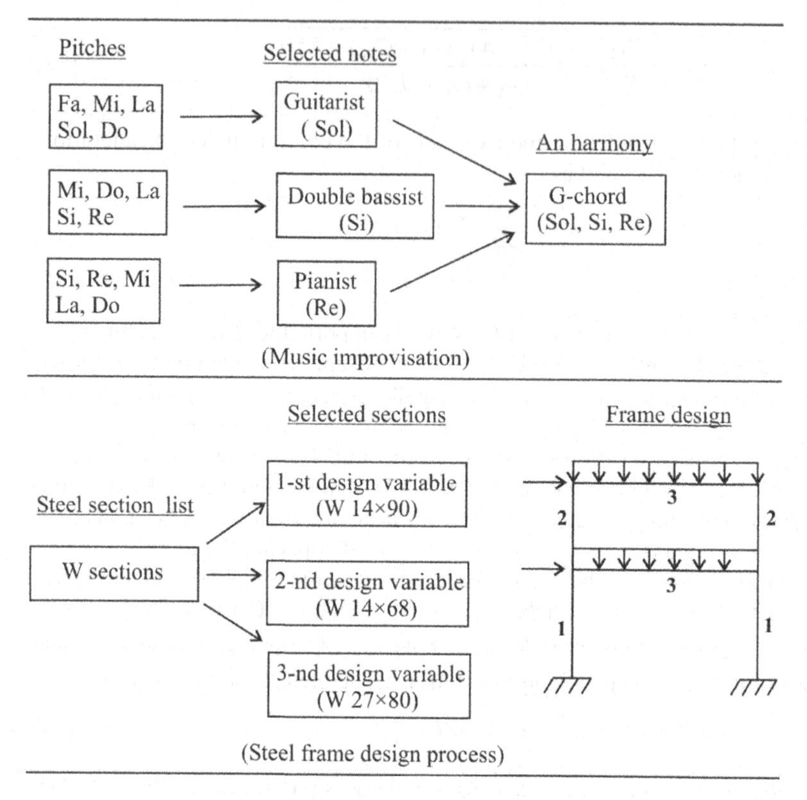

Fig. 2 Analogy between harmony memory and steel frame design process

better than existing worst harmony in the HM, new harmony is included in the HM and the existing worst harmony is excluded from the HM. The process is repeated until the best harmony is obtained.

We consider a steel frame design process, which consists of three different design variables. The first design variable is the columns of the first storey, the second design variable is the columns of the second storey and the third design variable is the all beams. The design variables are selected from a standard set of steel sections such as American Institute of Steel Construction (AISC) wide-flange (W) shapes. Let us assume W14x90, W14x68 and W27x80 are selected from the section list as the first, second and third design variables. Thus, a new steel design is created [W14x90, W14x68, W27x80]. If the new design is better than existing worst design which is the one with the highest objective function value, the new design is included and worst design is excluded from the steel design process. This procedure is repeated until terminating criterion is satisfied.

An analogy between the music improvisation process and the optimum design of steel frames can be established in the following way: The harmony denotes the design vector while the different harmonies during the improvisation represent the different design vectors throughout the optimum design process. Each musical instrument denotes the design variables (steel sections) of objective function. The pitches of the instruments represent the design variable's values (steel section no.). A better harmony represents local optimum and the best harmony is the global optimum.

4 Optimum Design Using Harmony Search Algorithm

The optimum design algorithm using HS could be illustrated as shown in Figure 3.

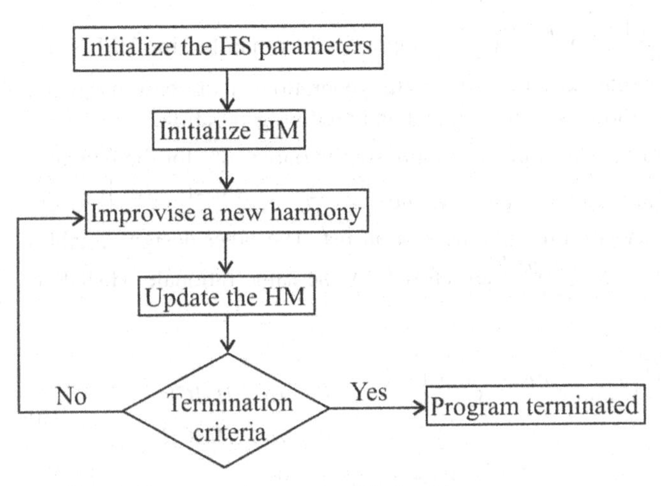

Fig. 3 Basic flowchart diagram for HS algorithm

4.1 Initialize the Harmony Search Parameters

The HS algorithm parameters are chosen in this step. These parameters are; harmony memory size (HMS), harmony memory consideration rate (HMCR), pitch adjusting rate (PAR) and stopping criteria (number of improvisation). They are selected depending on the problem type.

4.2 Initialize Harmony Memory

The harmony memory (HM) matrix is filled with randomly generated designs as the size of the harmony memory (HMS).

$$HM = \begin{bmatrix} x_1^1 & x_2^1 & \cdots & x_{ng-1}^1 & x_{ng}^1 \\ x_1^2 & x_2^2 & \cdots & x_{ng-1}^2 & x_{ng}^2 \\ \vdots & \vdots & \vdots & \vdots & \vdots \\ \vdots & \vdots & \vdots & \vdots & \vdots \\ x_1^{HMS-1} & x_2^{HMS-1} & \cdots & x_{ng-1}^{HMS-1} & x_{ng}^{HMS-1} \\ x_1^{HMS} & x_2^{HMS} & \cdots & x_{ng-1}^{HMS} & x_{ng}^{HMS} \end{bmatrix} \begin{matrix} \rightarrow & \varphi(x^1) \\ \rightarrow & \varphi(x^2) \\ \rightarrow & \vdots \\ \rightarrow & \vdots \\ \rightarrow & \varphi(x^{HMS-1}) \\ \rightarrow & \varphi(x^{HMS}) \end{matrix} \tag{15}$$

Each row represents a steel design in the HM. x^1, x^2,.....,x^{HMS-1}, x^{HMS} and $\varphi(x^1)$, $\varphi(x^2)$,..., $\varphi(x^{HMS-1})$, $\varphi(x^{HMS})$ are designs and the corresponding unconstrained objective function value, respectively. The steel designs in the HM are sorted by the unconstrained objective function values which are calculated by using Eqn. (2) (i.e. $\varphi(x^1)< \varphi(x^2)<......< \varphi(x^{HMS})$). The aim of using HM is to preserve better designs in the search process.

4.3 Improvise a New Harmony

A new harmony $[x^{nh}] = \lfloor x_1^{nh}, x_2^{nh}, ...x_{ng}^{nh} \rfloor$ is improvised from either the HM or entire section list. Three rules are applied for the generation of the new harmony. These are HM consideration, pitch adjustment and random generation.

In the HM consideration, the value of first design variable x_1^{nh} for the new harmony is chosen from any value in the HM (i.e. $\lfloor x_1^1, x_1^2, ..., x_1^{HMS-1}, x_1^{HMS} \rfloor$) or entire section list $[X_{SL}]$. $[X_{SL}]$ represents the section list. The other design variables of new harmony $\lfloor x_2^{nh}, ..., x_{ng-1}^{nh}, x_{ng}^{nh} \rfloor$ are chosen by the same rationale. HMCR is applied as follows

$$\begin{cases} x_i^{nh} \in \left\{ x_i^1, x_i^2, ..., x_i^{HMS-1}, x_i^{HMS} \right\} & \text{if} \quad rn \leq HMCR \\ x_i^{nh} \in X_{SL} & \text{if} \quad rn > HMCR \end{cases} \tag{16}$$

At first, a random number (rn) uniformly distributed over the interval [0,1] is generated. If this random number is equal or less than the HMCR value, i-th design variable of new design $[x^{nh}]$ selected from the current values stored in the i-th column of HM. If rn is higher than HMCR, i-th design variable of new design $[x^{nh}]$ is selected from the entire section list $[X_{SL}]$. For example, an HMCR of 0.90 shows that the algorithm will choose the i-th design variable (i.e. steel section) from the current stored steel sections in the i-th column of the HM with a 90% probability or from the entire section list with a 10% probability. A value of 1.0 for HMCR is not appropriate because of 0% possibility that the new design may be improved by values not stored in the HM [50].

Any design variable of the new harmony, $[x^{nh}] = \lfloor x_1^{nh}, x_2^{nh}, ...x_{ng}^{nh} \rfloor$, obtained by the memory consideration is examined to determine whether it is pitch-adjusted or not. Pitch adjustment is made by pitch adjustment ratio (PAR) which investigates better design in the neighbouring of the current design. PAR is applied as follows

$$\text{Pitch adjusting decision for } x_i^{nh} \leftarrow \begin{cases} yes & \text{if} \quad rna \leq PAR \\ no & \text{if} \quad rna > PAR \end{cases} \tag{17}$$

A random number (rna) uniformly distributed over the interval [0,1] is generated for x_i^{nh}. If this random number is less than the PAR, x_i^{nh} is replaced with its neighbour steel section in the section list. If this random number is not less than

PAR, x_i^{nh} remains the same. The selection of neighbour section is determined by neighbouring index. A PAR of 0.4 indicates that the algorithm chooses a neighbour section with a 40%×HMCR probability. For example, if x_i^{nh} is W21×62, neighbouring index is -1 or 1 and the section list is [W21×73, W21×68, *W21×62*, W21×57, W21×50], the algorithm will choose a neighbour section (W21×68 or W21×57) with a 40%×HMCR probability, or remain the same section (W21×62) with a (100%-40%)×HMCR probability. HMCR and PAR parameters are introduced to allow the solution to escape from local optima and to improve the global optimum prediction of the HS algorithm [50].

4.4 Update the Harmony Memory

If the new harmony $[x^{nh}] = \lfloor x_1^{nh}, x_2^{nh}, \ldots\ldots x_{ng}^{nh} \rfloor$ is better than the worst design in the HM (i.e. the last row of the HM), the new design is included in the HM and the existing worst harmony is excluded from the HM. In this process, it should be noted that HM matrix is sorted again by unconstrained objective function and the same design is not permitted in the HM more than once.

4.5 Termination Criteria

4.3 and 4.4 steps are repeated until the termination criterion is satisfied. In the present work, two termination criteria were used for HS. The first one stops the algorithm when a predetermined total number of searches (number of frame analyses) are performed. The second criterion stops the process before reaching the maximum search number, if more economical design (lighter frame) is not found during a definite number of searches in HS.

5 Optimum Design of Steel Frames Using Harmony Search Algorithm

The optimum design algorithm for steel frames using HS comprises the following steps.

Step1: Determine the harmony search parameters; HMS, HMCR, PAR and terminating criterion.

Step2: Generate a new design randomly for the steel frame. Analyze the frame for the new design and obtain its response. Calculate the value of the objective function $\varphi(x)$ using Eqn. (2). If it satisfies the constraints, record it as the optimum design $(\varphi(x)_{opt.})$. Repeat this step as many as 2×harmony memory (HM) matrix sizes. If any design in this process is a feasible one and better than the previous optimum (i.e. lower than $\varphi(x)_{opt}$) ,assign it as current optimum design. Sort the steel designs according to their $\varphi(x)$ values until harmony memory matrix is filled completely. The other designs out-of-HM are eliminated. The best design with the lowest $\varphi(x)$ one is denoted by $\varphi(x^{best})$ and placed in the first row of HM matrix and the worst design is denoted by $\varphi(x^{worst})$ and placed in the last row of HM matrix. The aim of generating 2×HM designs is to provide more appropriate initial designs for HM.

Step3: Select the first design variable of new design from either the first column of HM randomly or from steel section database according to HMCR value. If the design variable is selected from HM, decide whether to apply or not pitch-adjustment according to PAR parameter. Repeat this step until all design variables selected only once, and thus, obtain the new design vector for the steel frame. Analyze the frame for the new design vector and obtain its response. Calculate the value of the objective function $\varphi(x^{new})$ using Eqn. (2).

Step4: If $\varphi(x^{new})<\varphi(x^{worst})$, include new design in the HM and exclude the existing worst design from the HM. If the new design is also a feasible one and better than the previous optimum $\varphi(x)_{opt}$, assign it current optimum design. Sort the steel designs again in the HM according to their $\varphi(x)$ values.

Step5: Repeat 3-4 steps until the predetermined total number of searches (number of frame analyses) are performed or current optimum design remains the same during a definite number of searches. If one of these criteria is satisfied, terminate the algorithm and define the current optimum as the final optimum design.

6 Benchmark Examples

In this section, HS is utilized to the optimum design of five different steel frames, which are taken from current literature. The optimum designs of these frames were previously performed using GA, SA and ACO based algorithms [16,34,37]. They were designed again using HS algorithm and the design results were compared with the ones of the aforementioned algorithms. The presented examples in this section consist of author's previous studies [53,54].

6.1 Two-Bay, Three-Storey Planar Frame

The two-bay, three-storey frame under a single-load case shown in Figure 4 is the first benchmark example. This frame was optimized by Pezeshk et al. [16] using GA and it was also designed by Camp et al. [37] using ACO. Young's modulus of E=29,000 ksi and a yield stress of f_y=36 ksi were used.

Displacement constraints were not imposed for the design. The beam members were selected from a list with 267 W sections and W10 sections were used for column members. The member effective length factors K_x is calculated from the approximate equation proposed by Dumonteil [58]. For each column, the out-of-plane effective length factor (K_y) was considered as 1.0. The out-of-plane effective length factor for each beam member was specified to be 0.167. The penalty function exponent and penalty constant are taken as ε=2 and $\kappa = 1.0$ [37]. The size constraint given in Eqn. (7) was not imposed on the planar frame examples.

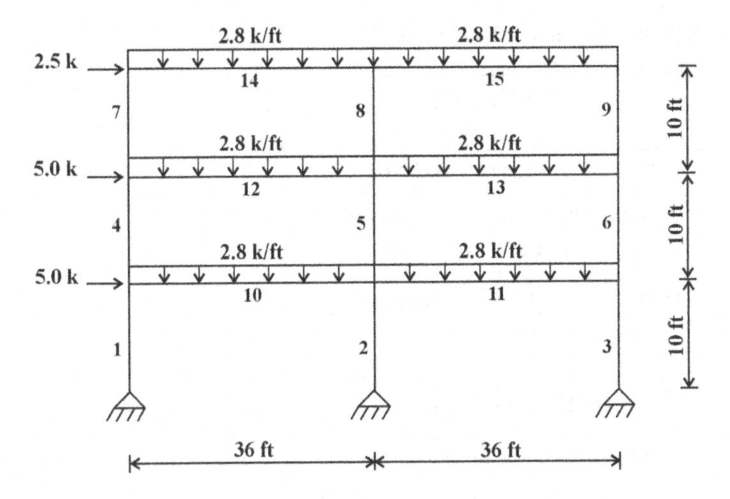

Fig. 4 Two-bay, three-storey frame

The HS algorithm, which was programmed in Fortran, is executed with the following tuning parameters: The harmony memory size (HMS) was selected depending on the design example. When HMS was selected greater than 100, HS did not improve the optimal solutions. For HMS<50, HS resulted in premature convergence. HMS is also sensitive to the number of design variables. When the number of design variables is increased, the search space enlarges. Therefore, larger HMS has to be used. HMS was selected as 50 in this example. Another tuning parameter affecting the results is the harmony memory consideration rate (HMCR), which was selected as 0.8. The higher values of HMCR tended to reach local optima, while the lower values of HMCR caused the non-optimal solutions. HS is also influenced by the value of pitch adjusting rate (PAR) which was taken as 0.4. Using higher values for PAR caused non-optimal designs, while lower values for it resulted in local optima. The neighboring index used in the pitch-adjustment selected as ±3. Lower values of ±3 caused the local optima, whereas higher values of ±3 diverged from the optimal designs. The maximum number of searches is another important parameter in the HS algorithm. Computational experience gained after different optimum designs shows that if the optimum design remains the same during the execution of 20% of the maximum search number, additional improvement is not made in the HS process. Therefore, the first and second termination criteria were selected as 5000 and 1000 in this example, respectively.

For HS algorithm, 30 different optimum frames were obtained generated from randomly selected 30 different initial designs. The lightest one of 30 optimum designs was assigned as the best optimum design and reported in Table 1. Typical design history for the best optimum design and average frame weight of 30 designs for two-bay, three-storey frame was also shown in Figure 5.

Table 1 Design results for two-bay, three-storey frame

Element group	GA [16]	ACO [37]	HS
Beam	W 24×62	W 24×62	W 21×62
Column	W 10×60	W 10×60	W 10×54
Weight (lb)	18,792	18,792	18,292
Number of analyses	1800	3000	1853

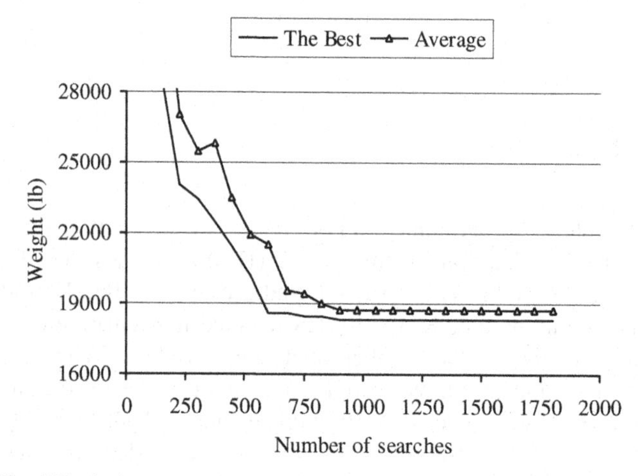

Fig. 5 Typical convergence history of two-bay three-storey frame

HS algorithm yielded 2.7% lighter frame than the ones obtained by GA and ACO. HS developed the best optimum design at the 853-th analysis and it did not change during 1000 frame analyses afterwards, and thus, HS terminated the search process after 1853 frame analyses. It was less than the 3000 frame analyses required by ACO, but it was more than the 1800 analyses required by GA.

The average weight of 30 different designs was calculated as 18,784 lb, with a standard deviation of 411 lb. HS developed the optimum designs with an average of 962-th frame analyses for 30 runs. The optimum designs were obtained within 900 and 880 frame analyses for GA and ACO, respectively. In this case, HS achieved lighter frame than the others with approximately the same analysis number.

It was an interesting result which was obtained from HS algorithm that interaction ratio of seven member is within 90% of maximum interaction ratio at optimum. This indicates that strength constraints were dominant at the optimum design. The maximum interaction ratio both column and beam group was calculated as 1.0 (i.e. boundary value).

6.2 One-Bay, Ten-Storey Planar Frame

The one-bay ten-storey planar frame is the second benchmark example. Figure 6 shows the frame configuration, the numbering of member groups and dimensions.

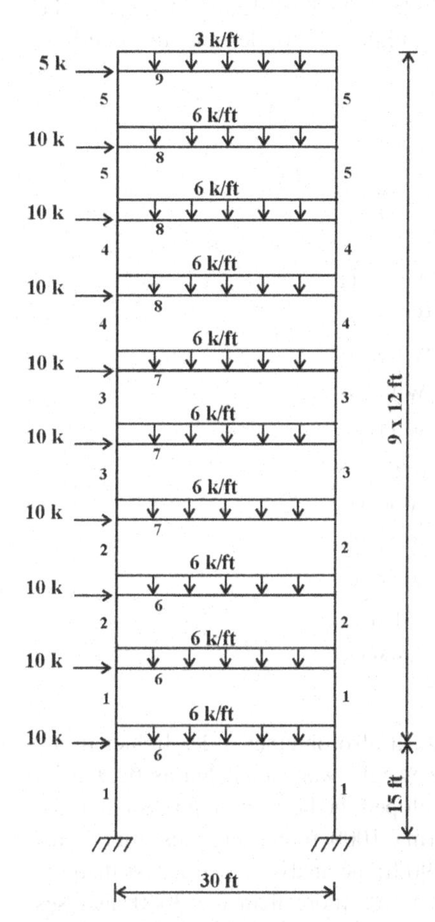

Fig. 6 One-Bay Ten-Storey Frame

This frame was optimized by Pezeshk et al. [16] using GA and then it was also optimized by Camp et al. [37] using ACO. The AISC-LRFD specification [56] was used and a displacement constraint was imposed as: interstorey drift<storey height/300. Young's modulus of E=29,000 ksi and a yield stress of fy=36 ksi were used. Beam element groups were chosen from 267 W-sections and five column groups were selected from only W14 and W12 sections. The effective length factors of members were calculated as $K_x \geq 1$ using the approximate equation proposed by Dumonteil [58], whereas the out-of-plane length factor K_y was assigned as 1. For each beam member, the out-of-plane effective length factor was

specified to be K_y=0.2 i.e., floor stringers at 1/5 points of the span. HMS was selected as 60 in this example. The other tuning parameters were the same as the first example.

30 different designs started with 30 different initial designs were performed and the lightest one of those was reported in Table 2. Typical design history for the best optimum design and average frame weight of 30 designs for one-bay, ten-storey frame was depicted in Figure 7.

Table 2 Design results for one-bay, ten-storey frame

Group no.	GA [16]	ACO [37]	HS
1	W 14×233	W 14×233	W 14×211
2	W 14×176	W 14×176	W 14×176
3	W 14×159	W 14×145	W 14×145
4	W 14×99	W 14×99	W 14×90
5	W 12×79	W 12×65	W 14×61
6	W 33×118	W 30×108	W 33×118
7	W 30×90	W 30×90	W 30×99
8	W 27×84	W 27 × 84	W 24×76
9	W 24×55	W 21×44	W 18×46
Weight (lb)	65,136	62,610	61,864
Number of analyses	3000	8300	3690

HS produced 5.0% lighter frame than GA. It also obtained 1.2% lighter frame than ACO. For 30 runs of HS, the average weight was calculated as 62,923 lb, with a standard deviation of 1.74 lb. HS developed the best optimum design at the 2690-th analysis and it did not change during 1000 frame analyses afterwards. Hence, the best optimum design required 3690 frame analyses. It was less than the 8300 analyses required by ACO, whereas it was more than the 3000 analyses required by GA.

HS developed the designs with an average of 2600-th frame analysis for 30 runs as shown in Figure 7. GA found the optimal designs within 40 generations with 2400 frame analyses while ACO required 6100 frame analyses. In this case, the average analysis number required by the HS is less than the one of ACO. It is slightly more than the one of GA.

Moreover; the interstorey drift constraint was within 90% of the upper limit between storeys 1 to 8 as shown in Figure 8. The interstorey drift was calculated as 0.48 in (i.e. boundary value) in the second, fourth and seventh storeys. This indicates that interstorey drift constraint was dominant at the optimum.

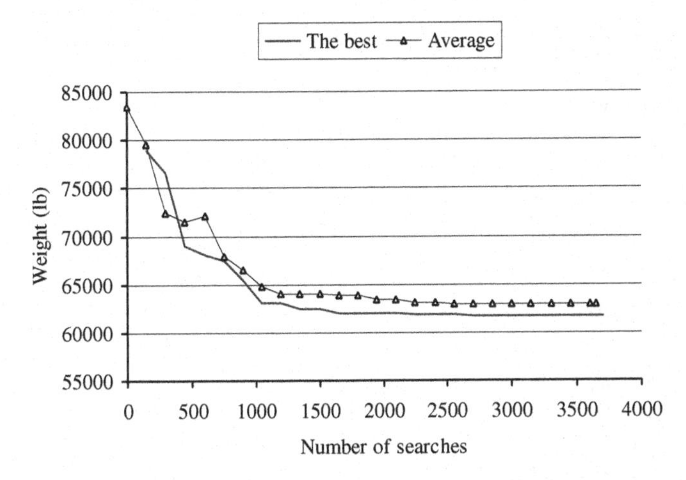

Fig. 7 Typical convergence history of one-bay ten-storey frame

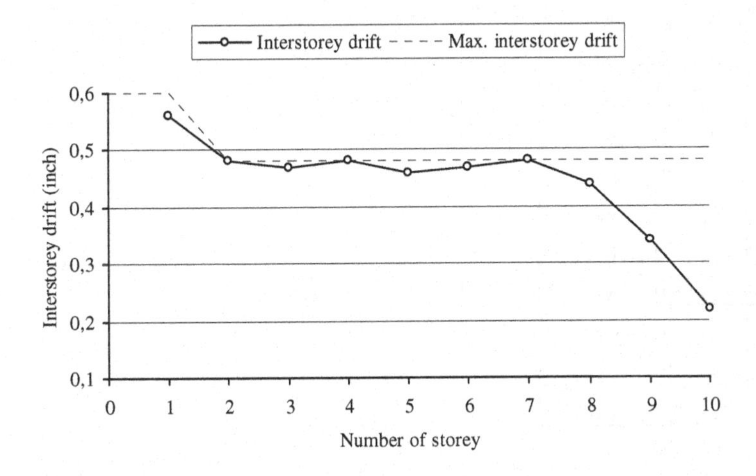

Fig. 8 Interstorey drift for one-bay ten-storey frame

6.3 Design of Three-Bay Twenty Four-Storey Planar Frame

The third example is the three-bay twenty-four storey steel frame shown in Figure 9.

This frame was designed by Camp et al [37] using ACO algorithm in accordance with AISC-LRFD [56]. The loads are W=5,761.85 lb, w_1=300 lb/ft, w_2=436 lb/ft, w_3=474 lb/ft and w_4=408 lb/ft. The frame was designed using AISC-LRFD [56] specification under the interstorey drift displacement constraint (interstorey drift<storey height/300). Young's modulus of E=29,732 ksi and a yield stress of f_y=33.4 ksi were used.

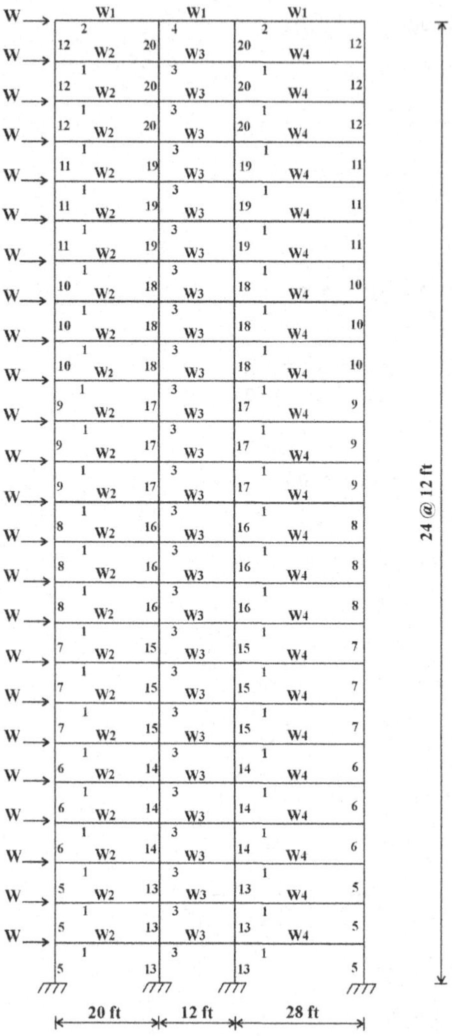

Fig. 9 Three-bay twenty four-storey frame

The effective length factors of the members are calculated as $K_x \geq 1.0$ from the approximate equation proposed by Dumonteil [58]. The out-of-plane effective length factor was K_y=1.0. All members were unbraced along their lengths. Each of the four beam element groups were chosen from all of the 267 W-sections, whereas the 16 column member groups were selected from only W14 sections. HMS was selected as 100 in this example. The first and second termination criteria were set 20000 and 4000.

For HS algorithm, 100 different designs started with different initial designs were executed and the lightest one of them was reported in Table 3. The average optimal weight for the 100 runs was 222,62 lb, a standard deviation of 5.8 lb.

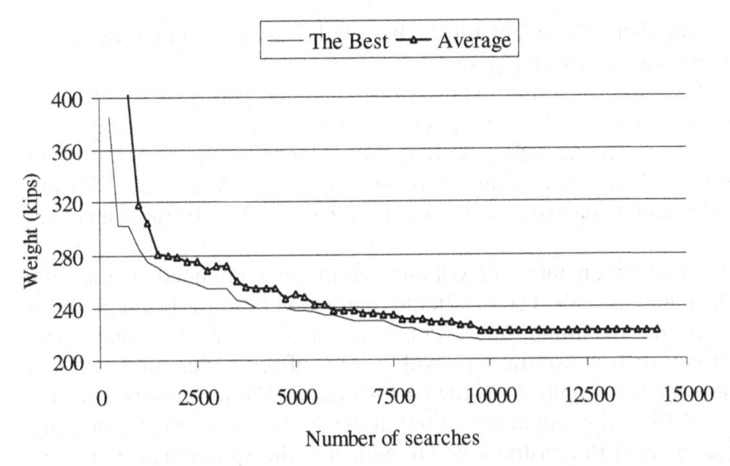

Fig. 10 Design history of three-bay twenty-four storey frame

Table 3 Design results for three-bay, twenty four-storey frame

Group no.	ACO [37]	HS
1	W 30×90	W 30×90
2	W 8×18	W 10×22
3	W 24×55	W 18×40
4	W 8×21	W 12×16
5	W 14×145	W 14×176
6	W 14×132	W 14×176
7	W 14×132	W 14×132
8	W 14×132	W 14×109
9	W 14×68	W 14×82
10	W 14×53	W 14×74
11	W 14×43	W 14×34
12	W 14×43	W 14×22
13	W 14×145	W 14×145
14	W 14×145	W 14×132
15	W 14×120	W 14×109
16	W 14×90	W 14×82
17	W 14×90	W 14×61
18	W 14×61	W 14×48
19	W 14×30	W 14×30
20	W 14×26	W 14×22
Weight (lb)	220,465	214,860

Design history of number of searches for the best and average optimum design of steel frame with HS was shown in Figure 10.

HS yielded 2.54% lighter frame than the one obtained using ACO. HS produced the best optimum at the 9924-th analysis and this design did not change during 4000 frame analyses afterwards, and thus, HS terminated the search process after 13924 frame analyses. According to the average of the results of 100 runs, HS required 14651 frame analyses, which was less than the 15500 frame analyses required by ACO.

The maximum interaction ratio for column groups was obtained in the 19th group. It was calculated as 0.89. The maximum interaction ratio for beam groups 1 and 3 was 0.78 and the maximum interaction ratio for beam groups 2 and 4 was 0.91. The interstorey drift constraint was within 90% of the upper limit between storeys 2 to 18 as shown in Figure 11 at the best optimum. The interstorey drift was also calculated as 0.48 in (i.e. boundary value) in the 11 storeys of the frame. This indicates that interstorey drift constraint was dominant in the optimum design.

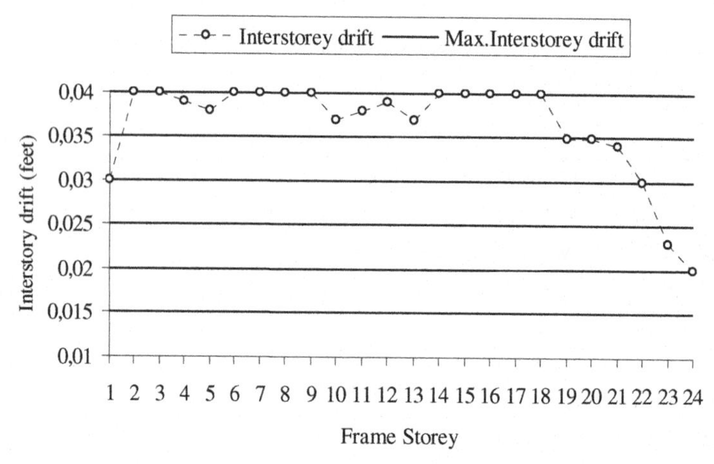

Fig. 11 Interstorey drift for three-bay twenty-four storey frame

6.4 Design of Single-Storey, 8-Member Space Frame

The single-storey, 8 member space frame shown in Figure 12 is the fourth benchmark example.

It was optimized by Degertekin [34] using SA and GA in accordance with the AISC-LRFD [56] and AISC-ASD [61] specifications. AISC-LRFD [56] yielded lighter frames than AISC-ASD [61]. This frame was optimized again using HS algorithm. The material properties were assigned as: a modulus of elasticity E=29000 ksi, a shear modulus of G=12000 ksi was used in the space frame structures. The yield stress and unit weight of material are 36 ksi and 0.2836 lb/in^3, respectively. Four different types of loads are employed: dead load (D), live load (L), roof live load (L_r), and wind loads (W). The following load combinations were used per AISC-LRFD specification: I: (1.4D), II: (1.2D + 1.6L + 0.5L_r), III: (1.2D + 1.6L_r + 0.5L), and IV: (1.2D + 1.3W + 0.5L + 0.5L_r).

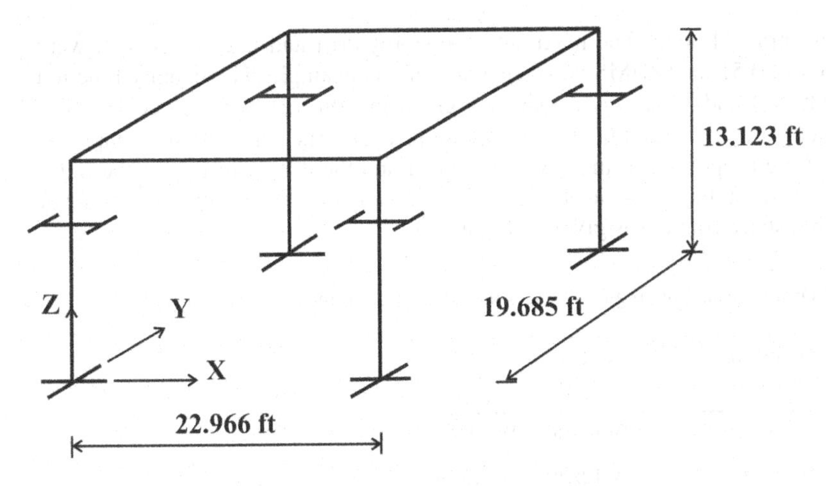

Fig. 12 Single-storey, 8-member space frame

The values of 0.464 psi for dead load (*D*), 0.348 psi for live load (*L*) and roof live load (L_r) were taken in the space frame structure. Wind loading was calculated from Uniform Building Code [62] using the equation $p = C_e C_q q_s I_w$, where p is design wind pressure; C_e is combined height, exposure and gust factor coefficient; C_q is pressure coefficient; q_s is wind stagnation pressure; and I_w is wind importance factor. Exposure D was assumed and the values for C_e were selected depending on the frame height and exposure type. The C_q values were assigned as 0.8 and 0.5 for inward and outward faces. The value of q_s was selected as 0.114 psi assuming a basic wind speed of 80 mph and the wind importance factor was assumed to be one. The horizontal loads due to wind act in the *x*-direction at each unrestrained node. The maximum drift of the top storey was restricted to *H*/400, where *H* is the total height of structure; the interstorey drift was also limited to h_c/300, where h_c is the height of the considered storey [63]. These limits were increased by 30% to include the effect of the coefficient 1.3 in the LRFD wind load combination.

Two section lists, comprised 64 W sections, were used for HS algorithm. The first one is beam section list taken from AISC-ASD [61]-Part 2, "Beam and Girder Design"- Allowable stress design selection table for shapes used as beams. The boldface type sections (lighter ones) were selected starting from W36×720 to W12×19. The second one is column section list taken from the same code, Part 3, "Column Design"- Column W shapes tables. They were selected from W14×283 to W6×15. The effective length factor *K*, for unbraced frames were calculated from the approximate equation proposed by Dumonteil [58]. Geometrically nonlinear analysis algorithm performed in HS is the same as SA algorithm [34]. Broad information about the geometrically nonlinear analysis algorithm can be found in the book by Levy and Spillers [64].

The members of the frame were divided into three groups organized as follows: 1-st group: the beams in *x*-direction, 2-nd group: the beams in *y*-direction, 3-rd group: the all columns. The horizontal loads due to wind act in the *x*-direction at each unrestrained node. The maximum top storey drift and interstorey drift were restricted to 0.51 in. 1×HMS was generated in this example. The penalty function exponent (ε) and penalty constant (κ) given in Eqn. (2) were selected as 1.0. The other parameters used in HS were the same as the ones of the first example.

10 different optimum frames were obtained and the design results of the lightest one were summarized in Table 4. The design history for the single-storey 8-member space frame was given in Figure 13.

Table 4 Design results of single-storey, 8-member space frame

Group no.	SA [34]	HS
1	W 12×30	W 12×26
2	W 12×30	W 12×26
3	W 8×24	W 8×28
Weight (lb)	3809	3693
Top storey drift (in)	0.49	0.50
Max. interstorey drift (in)	0.49	0.50
Number of analyses	6120	4412

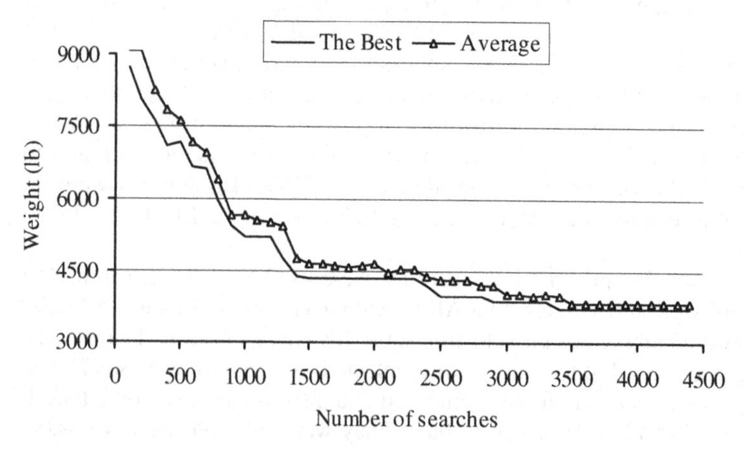

Fig. 13 Design history of the single-storey 8-member space frame

HS yielded 3.1% lighter frame than SA. HS produced the best optimum design after 4412 searches (i.e. 4412 frame analyses). This indicates that HS found the optimum design after 3412 searches and it did not change during 1000 searches afterwards. Both displacement and stress constraints are active in the best optimum design. The average weight of 10 runs was calculated as 3829.37 lb, with a standard deviation of 149.91 lb. HS converged at the optimum designs with an average of 4528 frame analyses for 10 runs while the optimum designs were obtained within 6120 frame analyses in SA. Therefore; HS obtained lighter frames than SA with less analyses' numbers.

6.5 Design of 4-Storey, 84-Member Space Frame

The last benchmark example is the 4-storey space frame with a square plan and side view shown in Figure 14. The structure consists of 84 members divided into 10 groups. It was designed by Degertekin [34] using SA. This frame designed again using HS in accordance with the AISC-LRFD [56].

The groups were organized as follows: 1-st group: outer beams of 4-th storey, 2-nd group: outer beams of 3-rd, 2-nd and 1-st storeys, 3-rd group: inner beams of 4-th storey, 4-th group: inner beams of 3-rd, 2-nd and 1-st storeys, 5-th group:

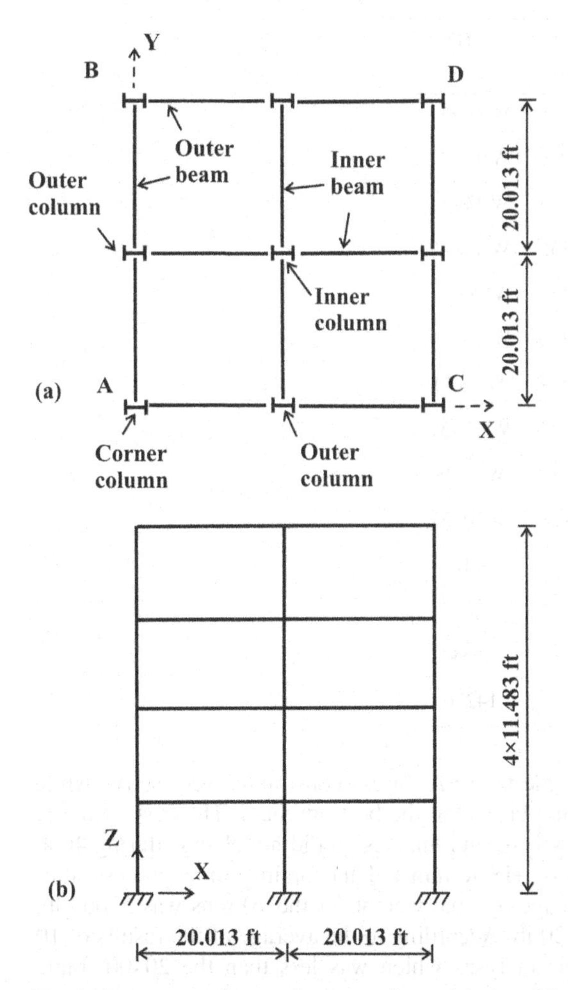

Fig. 14 Four-storey 84-member space frame (a) plan, (b) side view

corner columns of 4-th storey, 6-th group: corner columns of 3-rd, 2-nd and 1-st storeys, 7-th group: outer columns of 4-th storey, 8-th group: outer columns of 3-rd, 2-nd and 1-st storeys, 9-th group: inner columns of 4-th storey, 10-th group: inner columns of 3-rd, 2-nd and 1-st storeys. The wind loads act in the x-direction at each node on the sides AB and CD. For the maximum and interstorey drift constraints, the values of 1.79 in and 0.6 in were imposed on the frame. The first and second termination criteria were set 20000 and 4000. The other parameters were the same as the previous example.

For HS algorithms, 10 different designs were executed and design results of the lightest one of those were listed in Table 5. Design history for the 4-storey, 84-member space frame was shown in Figure 15.

Table 5 Design results of 4-storey 84-member space frame

Group no.	SA [34]	HS
1	W 18×35	W 16×31
2	W 18×35	W 16×31
3	W 18×35	W 16×31
4	W 18×35	W 16×40
5	W 8×31	W 8×31
6	W 12×40	W 10×39
7	W 10×39	W 8×40
8	W 12×45	W 10×39
9	W 8×28	W 8×28
10	W 12×58	W 10×77
Weight (lb)	50937	49019
Top storey drift (in)	1.74	1.69
Max. interstorey drift (in)	0.60	0.54
Number of analyses	20400	14276

HS obtained 3.8% lighter frame than SA. Stress constraints were active while displacement constraints were not critical at the best optimum. The best optimum was achieved at the 10276-th analysis and this design did not change during 4000 frame analyses afterwards. Thus, HS terminated the optimization process after 14276 frame analyses. The average optimal weight for the 10 runs was 51069 lb, with a standard deviation of 1420 lb. According to the average of the results of 10 runs, HS required 15532 frame analyses which was less than the 20400 frame analyses required by SA.

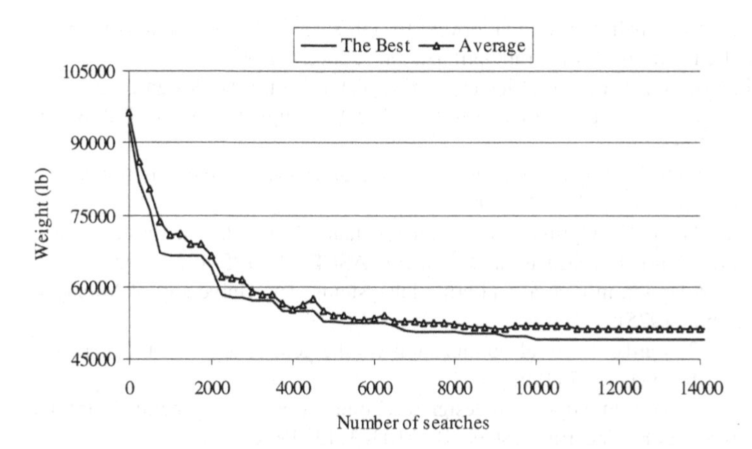

Fig. 15 Design history of the 4-storey 84-member space frame

7 Conclusions

A harmony search algorithm was applied to the optimum design of steel frames. The benchmark examples presented in this study revealed that HS is able to obtain lighter frames when compared to GA, SA and ACO. The reason for this is that HS provides a more effective and powerful approach than GA, SA and ACO as explained in Section 1. HS yielded 1.2%-5.0% lighter frames than the ones obtained from GA, ACO and SA based design. In addition to obtaining lighter frames, HS required less or approximately the same computational effort than GA, SA and ACO. The average weights of the frames in the examples were close to the optimum weights for HS. Standard deviations of the frames weights were also quite small in comparison with the frame weights. These prove that HS is able to find the global optima and it could be accepted as a powerful optimization technique for steel frame design using discrete and real design variables.

References

1. Arora, J.S.: Analysis of optimality criteria and gradient projection methods for optimal structural design. Computer Methods in Applied Mechanics and Engineering 23, 185–213 (1980)
2. Tabak, E.I., Wright, P.M.: Optimality criteria method for building frames. Journal of Structural Division ASCE 107, 1327–1342 (1981)
3. Lin, C.C., Liu, I.W.: Optimal design based on optimality criterion for frame structures including buckling constraints. Computers and Structures 31, 535–544 (1989)
4. Rozvany, G.I.N., Zhou, M.: A note on truss design for stress and displacement constraints by optimality criteria methods. Structural Optimization 3, 45–50 (1991)
5. Saka, M.P., Hayalioglu, M.S.: Optimum design of geometrically nonlinear elastic-plastic steel frames. Computers and Structures 38, 329–344 (1991)

6. Chan, C.M.: An optimality criteria algorithm for tall steel building design using commercial standard sections. Structural Optimization 5, 26–29 (1992)
7. Chan, C.M., Grierson, D.E.: An efficient resizing technique for the design of tall steel buildings subject to multiple drift constraints. The Structural Design of Tall Buildings 2, 17–32 (1993)
8. Soegiarso, R., Adeli, H.: Impact of vectorization on large scale structural optimization. Structural Optimization 7, 117–125 (1994)
9. Soegiarso, R., Adeli, H.: Optimum load and resistance factor design of steel space frame structures. Journal of Structural Engineering ASCE 123, 185–192 (1997)
10. Holland, J.H.: Adaption in natural and artificial systems. The Universtiy of Michigan Press, Ann Arbor (1975)
11. Jenkins, W.M.: Towards structural optimization via the genetic algorithm. Computers and Structures 40, 1321–1327 (1991)
12. Jenkins, W.M.: Plane frame optimum design environment based on genetic algorithm. Journal of Structural Engineering ASCE 118, 3103–3112 (1992)
13. Rajeev, S., Krishnamoorthy, C.S.: Discrete optimization of structures using genetic algorithms. Journal of Structural Engineering ASCE 118, 1233–1250 (1992)
14. Camp, C., Pezeshk, S., Cao, G.: Optimized design of two-dimensional structures using a genetic algorithm. Journal of Structural Engineering ASCE 124, 551–559 (1998)
15. Shrestha, S.M., Ghaboussi, J.: Evolution of optimum structural shapes using genetic algorithm. Journal of Structural Engineering ASCE 124, 1331–1338 (1998)
16. Pezeshk, S., Camp, C.V., Chen, D.: Design of nonlinear framed structures using genetic optimization. Journal of Structural Engineering ASCE 126, 382–388 (2000)
17. Hayalioglu, M.S.: Optimum design of geometrically non-linear elastic-plastic steel frames via genetic algorithm. Computers and Structures 77, 527–538 (2000)
18. Hayalioglu, M.S.: Optimum load and resistance factor design of steel space frames using genetic algorithm. Structural and Multidisciplinary Optimization 21, 292–299 (2001)
19. Toropov, V.V., Mahfouz, S.Y.: Design optimization of structural steelwork using a genetic algorithm, FEM and a system of design rules. Engineering Computations 18, 437–459 (2001)
20. Liu, M., Burns, S.A., Wen, Y.K.: Genetic algorithm based construction-conscious minimum weight design of seismic steel moment-resisting frames. Journal of Structural Engineering ASCE 132, 50–58 (2006)
21. Kameshki, E.S., Saka, M.P.: Optimum design of nonlinear steel frames with semi-rigid connections using a genetic algorithms. Computers and Structures 79, 1593–1604 (2001)
22. Kameski, E.S., Saka, M.P.: Genetic algorithm based optimum design of nonlinear planar steel frames with various semirigid connections. Journal of Constructional Steel Research 59, 109–134 (2003)
23. Hayalioglu, M.S., Degertekin, S.O.: Genetic algorithm based optimum design of nonlinear steel frames with semi-rigid connections. Steel and Composite Structures 4, 453–469 (2004)
24. Hayalioglu, M.S., Degertekin, S.O.: Design of non-linear steel frames for stress and displacement constraints with semi-rigid connections via genetic optimization. Structural and Multidisciplinary Optimization 27, 259–271 (2004)
25. Hayalioglu, M.S., Degertekin, S.O.: Minimum cost design of steel frames with semi-rigid connections and column bases via genetic optimization. Computers and Structures 83, 849–1863 (2005)

26. Metropolis, N., Rosenbluth, A., Teller, A., Teller, E.: Equation of state calculations by fast computing machines. Journal of Chemical Physics 21, 1087–1092 (1953)
27. Kirkpatrick, S., Gelatt, C.D., Vecchi, M.P.: Optimization by simulated annealing. Science 220, 671–680 (1983)
28. van Laarhoven, P.J.M., Aarts, E.H.L.: Simulated annealing: Theory and Applications. D. Riedel Publishing Company, Dordrecht (1987)
29. Balling, R.J.: Optimal steel frame design by simulated annealing. Journal of Structural Engineering ASCE 117, 1780–1795 (1991)
30. May, S.A., Balling, R.J.: A filtered simulated annealing strategy for discrete optimization of 3D steel frameworks. Structural Optimization 4, 142–148 (1992)
31. Huang, M.W., Arora, J.S.: Optimal design steel structures using standard sections. Structural Optimization 14, 24–35 (1997)
32. Park, H.S., Sung, C.W.: Optimization of steel structures using distributed simulated annealing algorithm on a cluster of personal computers. Computers and Structures 80, 1305–1316 (2002)
33. Degertekin, S.O.: A comparison of simulated annealing and genetic algorithm for optimum design of nonlinear steel space frames. Structural and Multidisciplinary Optimization 34, 347–359 (2007)
34. Hayalioglu, M.S., Degertekin, S.O.: Minimum-weight design of non-linear steel frames using combinatorial optimization algorithms. Steel and Composite Structures 7, 201–217 (2007)
35. Dorigo, M., Maniezzo, V., Colorni, A.: An investigation of some properties of an ant algorithm. In: Proceedings of the Parallel Problem Solving from Nature Conference (PPSN 1992), Brussels, Belgium (1992)
36. Camp, C.V., Bichon, B.J.: Design of space trusses using ant colony optimization. Journal of Structural Engineering ASCE 130, 741–751 (2004)
37. Camp, C.V., Bichon, B.J., Stovall, P.S.: Design of steel frames using ant colony optimization. Journal of Structural Engineering ASCE 131, 369–379 (2005)
38. Kaveh, A., Shojaee, S.: Optimal design of skeletal structures using ant colony optimization. International Journal for Numerical Methods in Engineering 70, 563–581 (2007)
39. Geem, Z.W., Kim, J.H., Loganathan, G.V.: A new heuristic optimization algorithm: harmony search. Simulation 76, 60–68 (2001)
40. Kim, J.H., Geem, Z.W., Kim, E.S.: Parameter estimation of the nonlinear muskingum model using harmony search. Journal of American Water Resources Association 37, 1131–1138 (2001)
41. Paik, K., Kim, J.H., Kim, H.S., Lee, D.R.: A conceptual rainfall-runoff model considering seasonal variation. Hydrological Processes 19, 3837–3850 (2005)
42. Geem, Z.W., Lee, K.S., Park, Y.: Application of harmony search to vehicle routing. American Journal of Applied Sciences 2, 1552–1557 (2005)
43. Geem, Z.W.: Optimal cost design of water distribution networks using harmony search. Engineering Optimization 38, 259–280 (2006)
44. Geem, Z.W.: Optimal scheduling of multiple dam system using harmony search algorithm. In: Sandoval, F., Prieto, A.G., Cabestany, J., Graña, M. (eds.) IWANN 2007. LNCS, vol. 4507, pp. 316–323. Springer, Heidelberg (2007)
45. Geem, Z.W., Williams, J.C.: Harmony search and ecological optimization. International Journal of Energy and Environment 1, 150–154 (2007)

46. Forsati, R., Haghighat, A.T., Mahdavi, M.: Harmony search based algorithms for bandwidth-delay-constrained least-cost multicast routing. Computer Communications 31, 2505–2519 (2008)
47. Cheng, Y.M., Li, L., Lansivaara, T., Chi, S.C., Sun, Y.J.: An improved harmony search minimization algorithm using different slip surface generation methods for slope stability analysis. Engineering Optimization 40, 95–115 (2008)
48. Lee, K.S., Geem, Z.W.: A new structural optimization method based on the harmony search algorithm. Computers and Structures 82, 781–798 (2004)
49. Lee, K.S., Geem, Z.W., Lee, S.H., Bae, K.W.: The harmony search heuristic algorithm for discrete structural optimization. Engineering Optimization 37, 663–684 (2005)
50. Lee, K.S., Geem, Z.W.: A new meta-heuristic algorithm for continuous engineering optimization: harmony search theory and practice. Computer Methods in Applied Mechanics and Engineering 194, 3902–3933 (2005)
51. Saka, M.P.: Optimum geometry design of geodesic domes using harmony search algorithm. Advances in Structural Engineering 10, 595–606 (2007)
52. Saka, M.P.: Optimum design of steel sway frames to BS 5950 using harmony search algorithm. Journal of Constructional Steel Research 65, 36–43 (2009)
53. Degertekin, S.O.: Harmony search algorithm for optimum design of steel frame structures: a comparative study with other optimization methods. Structural Engineering and Mechanics 29, 391–410 (2008)
54. Degertekin, S.O.: Optimum design of steel frames using harmony search algorithm. Structural and Multidisciplinary Optimization 36, 393–401 (2008)
55. Goldberg, D.E.: Genetic algorithms in search, optimization and machine learning. Addisson-Wesley, Reading (1989)
56. American Institute of Steel Construction, Manual of steel construction: load and resistance factor design. Chicago, Illionis (2001)
57. Arora, J.S.: Introduction to optimum design. Mc-Graw-Hill, New York (1989)
58. Dumonteil, P.: Simple equations for effective length factors. Engineering Journal AISC 29, 111–115 (1992)
59. Gaylord, E.H., Gaylord, C.N., Stallmeyer, J.E.: Design of steel structures. McGraw-Hill, New York (1992)
60. Galambos, T.V., Lin, F.J., Johnston, B.G.: Basic steel design with LRFD. Prentice Hall, New Jersey (1996)
61. American Institute of Steel Construction, Manual of steel construction: allowable stress design. Chicago, Illionis (1989)
62. Uniform Building Code, International Conference of Building Officials, Whittier, California (1997)
63. Ad Hoc Committee on Serviceability Research: Structural serviceability: a critical appraisal and research needs. Journal of Structural Engineering ASCE 112, 2646–2664 (1986)
64. Levy, R., Spillers, W.R.: Analysis of geometrically nonlinear structures. Chapman and Hall, New York (1994)

Adaptive Harmony Search Algorithm for Design Code Optimization of Steel Structures

M.P. Saka[1] and O. Hasançebi[2]

Abstract. In this chapter an improved version of harmony search algorithm called an adaptive harmony search algorithm is presented. The harmony memory considering rate and pitch adjusting rate are conceived as the two main parameters of the technique for generating new solution vectors. In the standard implementation of the technique, appropriate constant values are assigned to these parameters following a sensitivity analysis for each problem considered. The success of the optimization process is directly related to a chosen parameter value set. The adaptive harmony search algorithm proposed here incorporates a novel approach for adjusting these parameters automatically during the search for the most efficient optimization process. The efficiency of the proposed algorithm is numerically investigated using number of steel frameworks that are designed for minimum weight according to the provisions of various international steel design code specifications. The solutions obtained are compared with those of the standard algorithm as well as those of the other metaheuristic search techniques. It is shown that the proposed algorithm improves performance of the technique and it renders unnecessary the initial selection of the harmony search parameters.

1 Introduction

Design optimization of steel structures is important for structural engineers in today's world due to the fact that while the human population is increasing exponentially, the world resources are diminishing rapidly. More shelters are required to be built for living and more buildings are necessary to be constructed for production. Hence it is of the most importance that structures be designed and constructed by using minimum amount of material available. Optimum structural design algorithms provide a useful tool to steel designers to achieve this goal. These algorithms can be used to design a steel structure such that the design constraints specified by steel design codes are satisfied under the applied loads and the weight or the cost of the steel frame under consideration is the minimum. Formulation of the design optimization of steel structures produces a programming problem where the design variables are discrete in nature. The reason for this is that the steel sections to be adopted for frame members in practice are available

[1] Department of Engineering Sciences, Middle East Technical University, Ankara, Turkey
Email: mpsaka@metu.edu.tr
[2] Department of Civil Engineering, Middle East Technical University, Ankara, Turkey
Email: oguzhan@metu.edu.tr

Z.W. Geem (Ed.): Harmony Search Algo. for Structural Design Optimization, SCI 239, pp. 79–120.
springerlink.com

from a discrete list. Designer has the option of assigning any one of these available steel sections from the list to any one of member groups in the frame either arbitrarily or using his or her previous experience. Once an assignment is carried out for all the member groups in the frame, designer has a candidate solution in her or his hand for the design problem. It then becomes necessary to analyze the frame with these selected steel profiles to find out whether the response of the frame under the external loading is within the limitations set by design codes. If the result of analysis reveals that these limitations are satisfied then the designer has a feasible solution to the frame design problem. It is quite natural that the designer wonders whether there are other solutions would require less steel. As a result of this curiosity the search continues until the designer locates a feasible design which is better then the previously obtained designs in terms of the material required for its construction. It is apparent that this procedure is quite time consuming because quite large number of combinations is possible for the member groups of a frame depending upon the total number of practically available steel sections. For example for a frame where the members are collected in nine groups and that the total number of available steel profile sections is 120, there are 5.16×10^{18} possible combinations each of which can be a possible candidate for the frame under consideration and required to be tried. Some of these combinations may be eliminated by making use of designer's practical experience but still checking the remaining possibilities needs enormous computation time and effort to locate the optimum combination of steel sections. It is apparent that practicing structural designer will have neither time nor resources to carry out this search which covers all the possibilities. He or she will take the decision about steel sections to be used for member groups after few trials. Hence one of the feasible s olutions will be used for the design but not the optimum one.

Obtaining the solution of combinatorial optimization problems described above is not an easy task. Until recently the numbers of solution techniques available in the literature that can be used to determine the optimum solution of discrete programming problems were limited and their efficiency in large size design problems was challenging [1,2]. The emergence of meta-heuristic optimization techniques has opened a new era in obtaining the solution of such programming problems [3-7]. These techniques make use of ideas taken from the nature such as survival of the fittest, immune system or cooling of molten metals through annealing to develop a numerical optimization algorithm. These methods are non-traditional stochastic search and optimization methods and they are very suitable and effective in finding the solution of combinatorial optimization problems. They do not require the gradient information of the objective function and constraints and they use probabilistic transition rules not deterministic rules. They are shown to be quite effective in finding the optimum solution of optimization problems where the design variables are discrete. Among those available in the literature are simulated annealing, evolution strategies, particle swarm optimizer, tabu search method, genetic algorithm and ant colony optimization. As can be understood from their names each technique simulates one particular phenomenon that exists in the nature. There are large numbers of structural optimization procedures available in the literature each is based one of these techniques.

Harmony search algorithm is a new addition to this category of numerical optimization procedures and it simulates a jazz musician's improvisation [8-10]. It resembles an analogy between the attempt to find the harmony in music and the effort to find the optimum solution of an optimization problem. As the aim of a musician is to attain a piece of music with perfect harmony, the task of an optimizer is to come up with the optimum solution that satisfies all the constraints in the problem and minimizes the objective function. Naturally certain rules and parameters are used to transfer this innovative thinking into a numerical optimization technique. For example when a musician is improvising there are three possibilities. A tune can be played from musician's memory or above mentioned tune can be pitch adjusted or a tune can be played totally randomly. Harmony search method is based on these options. It may randomly select a steel section within previously identified and collected group of feasible sections, it may or may not apply pitch adjustment to this section depending on some random rule, or a steel section may randomly be selected from the entire steel sections list. The collected group of feasible solutions is stored in harmony memory matrix. Harmony search method is applied to various structural design optimization problems and found to be quite effective in obtaining their solution.

In this chapter code based design optimization of steel frames is first presented. The mathematical modeling of the discrete optimum design problem of steel frames formulated according to the provisions of Allowable Stress Design code of American Institute of Steel Construction (ASD-AISC) [11], Load and Resistance Factor Design (LRFD-AISC) [12] and British Steel design Code (BS 5950) [13] are described. This is followed by the presentation of the adaptive harmony search method which is an improved version of harmony search algorithm. In this technique two main parameters of the standard harmony search technique that is the harmony memory considering rate and pitch adjusting rate are adjusted automatically during the search procedure. In the standard implementation of the technique appropriate constant values are assigned to these parameters following a sensitivity analysis for each problem considered. The success of the optimization process is directly related to a chosen parameter value set. The adaptive harmony search algorithm presented in this chapter adjusts these parameters automatically during the search for the most efficient optimization process. The efficiency of the adaptive harmony search technique is numerically investigated by considering three design optimization problems of steel frames. The first one is three dimensional 209-member industrial steel frame. The second one is the three dimensional 568-member moment resisting steel frame. The third one is the 1890-member three dimensional braced steel frame. All these frames are designed for minimum weight according to provisions of Allowable Stress Design Code of American Institute of Steel Construction (ASD-AISC). The solutions obtained are compared with those of the standard harmony search algorithm as well as of the other metaheuristic search techniques. It is apparent that the design examples are selected among the real size steel frames that can be found in practice. In the following section the optimum design of the 115-member braced plane frame is carried out according to various international design code specifications namely, ASD-AISC, LRFD-AISC and BS 5950 using adaptive harmony search technique. The results obtained are compared to demonstrate the relationship between the design code used and the optimum solution obtained.

2 Code Based Design Optimization of Steel Frames

The formulation of the design optimization problem of a steel frame according to a steel design code yields itself to a discrete programming problem, if steel profiles for its members are to be selected from available steel sections list. The mathematical model of the design optimization problems depending on three international steel design codes considered in the formulation is described in the following.

2.1 Discrete Optimum Design of Steel Frames to ADS-AISC

Consider a steel structure consisting of *nm* members that are collected in *ng* design groups (variables). If the provisions of ASD-AISC [11] code are to be used in the formulation of the design optimization problem and the design groups are selected from given steel sections profile list, the following discrete programming problem is obtained.

Find a vector of integer values **I** (Eqn. 1) representing the sequence numbers of steel sections assigned to *ng* member groups

$$\mathbf{I}^T = \left[I_1, I_2, \ldots, I_{ng}\right] \tag{1}$$

to minimize the weight (*W*) of the frame

$$W = \sum_{i=1}^{ng} m_i \sum_{j=1}^{nt} L_j \tag{2}$$

where m_i is the unit weight of the steel section adopted for member group *i*, respectively, *nt* is the total number of members in group *i*, and L_j is the length of the member *j* which belongs to group *i*.

The members subjected to a combination of axial compression and flexural stress must be sized to meet the following stress constraints:

$$if\ \frac{f_a}{F_a} > 0.15; \quad \left[\frac{f_a}{F_a} + \frac{C_{mx} f_{bx}}{\left(1 - \frac{f_a}{F'_{ex}}\right) F_{bx}} + \frac{C_{my} f_{by}}{\left(1 - \frac{f_a}{F'_{ey}}\right) F_{by}}\right] - 1.0 \le 0 \tag{3}$$

$$\left[\frac{f_a}{0.60 F_y} + \frac{f_{bx}}{F_{bx}} + \frac{f_{by}}{F_{by}}\right] - 1.0 \le 0 \tag{4}$$

$$if \frac{f_a}{F_a} \leq 0.15; \quad \left[\frac{f_a}{F_a} + \frac{f_{bx}}{F_{bx}} + \frac{f_{by}}{F_{by}} \right] - 1.0 \leq 0 \tag{5}$$

If the flexural member is under tension, then the following formula is used instead:

$$\left[\frac{f_a}{0.60F_y} + \frac{f_{bx}}{F_{bx}} + \frac{f_{by}}{F_{by}} \right] - 1.0 \leq 0 \tag{6}$$

In Eqns. (3-6), F_y is the material yield stress, and $f_a = (P / A)$ represents the computed axial stress, where A is the cross-sectional area of the member. The computed flexural stresses due to bending of the member about its major (x) and minor (y) principal axes are denoted by f_{bx} and f_{by}, respectively. F'_{ex} and F'_{ey} denote the Euler stresses about principal axes of the member that are divided by a factory of safety of 23/12. F_a stands for the allowable axial stress under axial compression force alone, and is calculated depending on elastic or inelastic buckling failure mode of the member using Formulas 1.5-1 and 1.5-2 given in ASD-AISC [11]. For an axially loaded bracing member whose slenderness ratio exceeds 120, F_a is increased by a factor of $(1.6 - L / 200r)$ considering relative unimportance of the member, where L and r are the length and radii of gyration of the member, respectively. The allowable bending compressive stresses about major and minor axes are designated by F_{bx} and F_{by}, which are computed using the Formulas 1.5-6a or 1.5-6b and 1.5-7 given in ASD-AISC [11]. C_{mx} and C_{my} are the reduction factors, introduced to counterbalance overestimation of the effect of secondary moments by the amplification factors $(1 - f_a / F'_e)$. For unbraced frame members, they are taken as 0.85. For braced frame members without transverse loading between their ends, they are calculated from $C_m = 0.6 - 0.4(M_1 / M_2)$, where M_1 / M_2 is the ratio of smaller end moment to the larger end moment. Finally, for braced frame members having transverse loading between their ends, they are determined from the formula $C_m = 1 + \psi(f_a / F'_e)$ based on a rational approximate analysis outlined in ASD-AISC [11] Commentary-H1, where ψ is a parameter that considers maximum deflection and maximum moment in the member.

For computation of allowable compression and Euler stresses, the effective length factors K are required. For beam and bracing members, K is taken equal to unity. For column members, alignment charts are furnished in ASD-AISC [11] for calculation of K values for both braced and unbraced cases. In this study, however, the following approximate effective length formulas are used based on Dumonteil [14], which are accurate within about -1.0 and +2.0 % of exact results [15]:

For unbraced members:

$$K = \sqrt{\frac{1.6G_A G_B + 4(G_A + G_B) + 7.5}{G_A + G_B + 7.5}} \tag{7}$$

For braced members:

$$K = \frac{3G_A G_B + 1.4(G_A + G_B) + 0.64}{3G_A G_B + 2.0(G_A + G_B) + 1.28} \tag{8}$$

where G_A and G_B refer to stiffness ratio or relative stiffness of a column at its two ends.

It is also required that computed shear stresses (f_v) in members are smaller than allowable shear stresses (F_v), as formulated in Eqn. (9).

$$f_v \leq F_v = 0.40 C_v F_y \tag{9}$$

In Eqn. (9), C_v is referred to as web shear coefficient. It is taken equal to $C_v = 1.0$ for rolled I-shaped members with $h / t_w \leq 2.24E / F_y$, where h is the clear distance between flanges, E is the elasticity modulus and t_w is the thickness of web. For all other symmetric shapes, C_v is calculated from Formulas G2-3, G2-4 and G2-5 in ANSI/AISC 360-05 [16].

Apart from stress constraints, slenderness limitations are also imposed on all members such that maximum slenderness ratio ($\lambda = KL / r$) is limited to 300 for members under tension, and to 200 for members under compression loads. The displacement constraints are imposed such that the maximum lateral displacements are restricted to be less than H/400, and upper limit of story drift is set to be h/400, where H is the total height of the frame building and h is the height of a story.

Finally, we consider geometric constraints between beams and columns framing into each other at a common joint for practicality of an optimum solution generated. For the two beams B1 and B2 and the column shown in Figure 1, one can write the following geometric constraints:

$$\frac{b_{fb}}{b_{fc}} - 1.0 \leq 0 \tag{10}$$

$$\frac{b'_{fb}}{(d_c - 2t_f)} - 1.0 \leq 0 \tag{11}$$

where b_{fb}, b'_{fb} and b_{fc} are the flange width of the beam B1, the beam B2 and the column, respectively, d_c is the depth of the column, and t_f is the flange width of the column. Equation (10) simply ensures that the flange width of the beam B1 remains smaller than that of the column. On the other hand, Eqn. (11) enables that

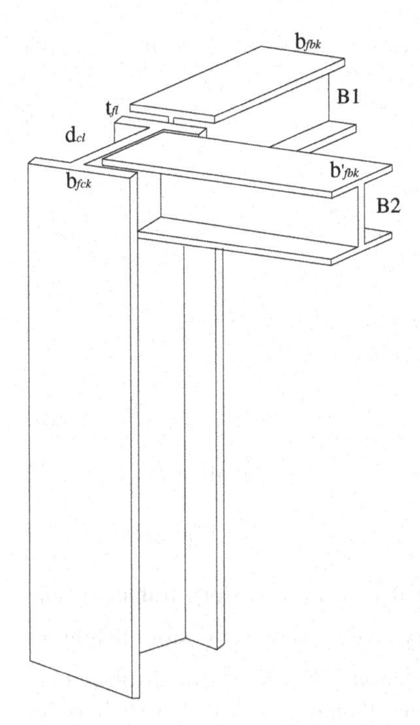

Fig. 1 Beam-column geometric constraints.

flange width of the beam B2 remains smaller than clear distance between the flanges of the column $(d_c - 2t_f)$.

2.2 Discrete Optimum Design of Steel Frames to LRFD-AISC

In the case where the optimum design problem of a steel frame is formulated according to the provisions of LRFD-AISC [12] the following discrete programming problem is obtained.

Find a vector of integer values **I** (Eqn. 12) representing the sequence numbers of steel sections assigned to *ng* member groups

$$\mathbf{I}^T = [I_1, I_2, ..., I_{ng}] \tag{12}$$

to minimize the weight (*W*) of the frame

$$W = \sum_{i=1}^{ng} m_i \sum_{j=1}^{nt} L_j \tag{13}$$

where m_i is the unit weight of the steel section adopted for member group *i*, respectively, *nt* is the total number of members in group *i*, and L_j is the length of the member *j* which belongs to group *i*. The following constraints are required to be imposed according to LRFD-AISC provisions.

$$(\delta_j - \delta_{j-1})/h_j \leq \delta_{ju} \qquad j = 1,\ldots, ns \qquad (14)$$

$$\delta_i \leq \delta_{iu} \qquad i = 1,\ldots, nd \qquad (15)$$

$$\frac{P_{uk}}{\phi P_{nk}} + \frac{8}{9}\left(\frac{M_{uxk}}{\phi_b M_{nxk}} + \frac{M_{uyk}}{\phi_b M_{nyk}}\right) \leq 1 \quad for \frac{P_{uk}}{\phi P_{nk}} \geq 0.2 \qquad k = 1,\ldots, nc \qquad (16)$$

$$\frac{P_{uk}}{2\phi P_{nk}} + \left(\frac{M_{uxk}}{\phi_b M_{nxk}} + \frac{M_{uyk}}{\phi_b M_{nyk}}\right) \leq 1 \quad for \frac{P_{uk}}{\phi P_{nk}} < 0.2 \qquad k = 1,\ldots, nc \qquad (17)$$

$$M_{uxt} \leq \phi M_{nxt} \qquad t = 1,\ldots\ldots, nb \qquad (18)$$

$$b_{fb} \leq b_{fc} \qquad \text{at all joints} \qquad (19)$$

$$b_{fb} \leq \left(d_c - 2t_f\right) \qquad \text{at all joints} \qquad (20)$$

where Eqn. (14) represents the inter-story drift of the multi-story frame. δ_j and δ_{j-1} are lateral deflections of two adjacent story levels and h_j is the story height. *ns* is the total number of storys in the frame. Equation (15) defines the displacement restrictions that may be required to include other than drift constraints such as deflections in beams. *nd* is the total number of restricted displacements in the frame. δ_{ju} is the allowable lateral displacement. The allowable lateral displacements are restricted to be less than *H*/400, and upper limit of story drift is set to be *h*/400, where *H* is the total height of the frame building and *h* is the height of a story.

Eqns. (16) and (17) represent strength constraints for doubly and singly symmetric steel members subjected to axial force and bending. If the axial force in member *k* is tensile force, the terms in these equations are given as: P_{uk} is the required axial tensile strength, P_{nk} is the nominal tensile strength, ϕ becomes ϕ_t in the case of tension and called strength reduction factor which is given as 0.90 for yielding in the gross section and 0.75 for fracture in the net section, ϕ_b is the strength reduction factor for flexure given as 0.90, M_{uxk} and M_{uyk} are the required flexural strength, M_{nxk} and M_{nyk} are the nominal flexural strength about major and minor axis of member *k* respectively. It should be pointed out that required flexural bending moment should include second-order effects. LRFD suggests an approximate procedure for computation of such effects which is explained in C1 of LRFD. In the case the axial force in member *k* is compressive force, the terms in Eqns. (16) and (17) are defined as: P_{uk} is the required compressive strength, P_{nk} is the nominal compressive strength, and ϕ becomes ϕ_c which is the resistance factor for compression given as 0.85. The remaining notations in Eqns. (16) and (17) are the same as the definition given above.

The nominal tensile strength of member k for yielding in the gross section is computed as $P_{nk} = F_y A_{gk}$ where F_y is the specified yield stress and A_{gk} is the gross area of member k. The nominal compressive strength of member k is computed as $P_{nk} = A_{gk} F_{cr}$ where $F_{cr} = \left(0.658^{\lambda_c^2}\right) F_y$ for $\lambda_c \leq 1.5$ and $F_{cr} = \left(0.877 / \lambda_c^2\right) F_y$ for $\lambda_c > 1.5$ and $\lambda_c = \frac{Kl}{r\pi}\sqrt{F_y / E}$. In these expressions E is the modulus of elasticity, K and l are the effective length factor and the laterally unbraced length of member k respectively.

Equation (18) represents the strength requirements for beams in load and resistance factor design according to LRFD-F2. M_{uxt} and M_{nxt} are the required and the nominal moment about major axis in beam b respectively. ϕ_b is the resistance factor for flexure given as 0.90. M_{nxt} is equal to M_p, plastic moment strength of beam b which is computed as ZF_y where Z is the plastic modulus and F_y is the specified minimum yield stress for laterally supported beams with compact sections. The computation of M_{nxb} for non-compact and partially compact sections is given in Appendix F of LRFD.

Equation (19) is included in the design problem to ensure that the flange width of the beam section at each beam-column connection of story s should be less than or equal to the flange width of column section. Equation (20) enables that flange width of the beam B2 remains smaller than clear distance between the flanges of the column $(d_c - 2t_f)$. The notations in Eqns. (19) and (20) are shown in Figure 1.

2.3 Discrete Optimum Design of Steel Frames to BS5950

In case BS5950 [13] is used in formulation of the optimum design problem of a steel frame, the following discrete programming problem is obtained.

Find a vector of integer values **I** (Eqn. 21) representing the sequence numbers of steel sections assigned to ng member groups

$$\mathbf{I}^T = \left[I_1, I_2, \ldots, I_{ng}\right] \tag{21}$$

to minimize the weight (W) of the frame

$$W = \sum_{i=1}^{ng} m_i \sum_{j=1}^{nt} L_j \tag{22}$$

where m_i is the unit weight of the steel section adopted for member group i, respectively, nt is the total number of members in group i, and L_j is the length of the member j which belongs to group i. The following constraints are required to be imposed according to BS5950 provisions.

$$(\delta_j - \delta_{j-1}) / h_j \leq \delta_{ju} \qquad j = 1, \ldots, ns \tag{23}$$

$$\delta_i \le \delta_{iu} \qquad i = 1,...., nd \tag{24}$$

$$\frac{F_k}{A_{gk} p_y} + \frac{M_{xk}}{M_{cxk}} + \frac{M_{yk}}{M_{cyk}} \le 1 \qquad k = 1,...., nc \tag{25}$$

or

$$\frac{F_k}{A_{gk} P_{ck}} + \frac{m_k M_{xk}}{M_{bk}} + \frac{m_k M_{yk}}{p_y Z_{yk}} \le 1 \qquad k = 1,...., nc \tag{26}$$

$$M_{xn} \le M_{cxn} \qquad t = 1,......, nb \tag{27}$$

$$b_{fb} \le b_{fc} \qquad \text{at all joints} \tag{28}$$

$$b_{jb} \le (d_c - 2t_f) \qquad \text{at all joints} \tag{29}$$

Equation (23) represents the inter-story drift of the multi-story frame. δ_j and δ_{j-1} are lateral deflections of two adjacent story levels and h_j is the story height. *ns* is the total number of storys in the frame. Equation (24) defines the displacement restrictions that may be required to include other than drift constraints such as deflections in beams. *nd* is the total number of restricted displacements in the frame. δ_{ju} is the allowable lateral displacement. BS 5950 limits the horizontal deflection of columns due to unfactored imposed load and wind loads to height of column/300 in each story of a building with more than one story. δ_{iu} is the upper bound on the deflection of beams which is given as span/360 if they carry plaster or other brittle finish.

Equation (25) defines the local capacity check for beam-columns. F_k, M_{xk} and M_{yk} are the applied axial load and moments about the major and minor axis at the critical region of member k respectively. A_{gk} is the gross cross sectional area, and p_y is the design strength of the steel. M_{cxk} and M_{cyk} are the moment capacities about major and minor axis of member *k*. nc is the total number of beam-columns in the frame.

Equation (26) represents the simplified approach for the overall buckling check for beam-columns. m_k is the equivalent uniform moment factor of member k given in table 18 of BS 5950. M_{bk} is the buckling resistance moment capacity for member k about its major axis computed from clause 4.3.7 of the code. Z_{yk} is the elastic section modulus about the minor axis of member *k*. P_{ck} is the compression strength obtained from the solution of quadratic Perry-Robertson formula given in appendix C.1 of BS 5950. It is apparent that computation of the compressive strength of a compression member requires its effective length. This can be automated by using Jackson and Moreland monograph for frame buckling [17].

The relationship for the effective length of a column in a swaying frame is given as:

$$\frac{(\gamma_1\gamma_2)(\pi/k)^2-36}{6(\gamma_1+\gamma_2)}=\frac{\pi/k}{\tan(\pi/k)} \tag{30}$$

where k is the effective length factor and γ_1 and γ_2 are the relative stiffness ratio for the compression member which are given as:

$$\gamma_1=\frac{\sum I_{c1}/\ell_{c1}}{\sum I_{b1}/\ell_{b1}} \quad \text{and} \quad \gamma_2=\frac{\sum I_{c2}/\ell_{c2}}{\sum I_{b2}/\ell_{b2}} \tag{31}$$

The subscripts c and b refer to the compressed and restraining members respectively and the subscripts 1 and 2 refer to two ends of the compression member under investigation. The solution of the nonlinear equation (30) for k results in the effective length factor for the member under consideration. The Eqn. (30) has the following form for non-swaying frames.

$$\frac{\gamma_1\gamma_2}{4}\left(\frac{\pi}{k}\right)^2+\left(\frac{\gamma_1+\gamma_2}{2}\right)\left(1-\frac{\pi/k}{\tan(\pi/k)}\right)+\frac{2\tan(\pi/2k)}{\pi/k}=1 \tag{32}$$

The notations in the remaining inequalities (28) and (29) are the same as those defined in inequalities (19) and (20).

3 Adaptive Harmony Search Algorithm

Harmony search method is a recent meta-heuristic technique that is shown to be effective and robust in obtaining the optimum solution of discrete programming problems. Its use in structural optimization and computational mechanics is still new. Among the few numbers of studies Lee and Geem [9] applied the method to determine the optimum design of plane and space trusses with continuous design variables. The method is used in the optimum design of steel frames with discrete variables by Değertekin [18] and Saka [19] where the design problem is formulated according to LRFD-AISC and BS5950 respectively. Later the same technique is employed in the optimum design of grillage systems [20, 21]. It is shown by Saka [22, 23] that harmony search algorithm can also be used in shape optimization problems. In this study harmony search method has successfully determined the optimum height of a geodesic dome in addition to pipe section designations for its members. It is demonstrated within these studies that harmony search method was a rapid and effective method for optimum design of structural systems where the number of design variables was relatively small. However, a comprehensive performance evaluation of harmony search method carried out at Hasançebi et al. [24, 25] in real size large scale structural optimization problems has shown that this conclusion were only true for small size problems. In this study the technique is compared with other meta-heuristic algorithms and found out that in large scale design optimization problems the technique has demonstrated slow convergence rate and heavier optimum designs. Hence it became necessary to suggest some improvements in the standard harmony search method so that the above mentioned discrepancy can be eliminated and the method demonstrates similar performance with other meta-heuristic techniques in the case of large scale design problems. With this amendment an improved technique called adaptive harmony search method is formulated and proposed in Hasançebi et al. [26].

In standard harmony search method there are two parameters known as harmony memory considering rate ($hmcr$) and pitch adjusting rate (par) that play an important role in obtaining the optimum solution. These parameters are assigned to constant values that are arbitrarily chosen within their recommended ranges by Geem [27-29] based on the observed efficiency of the technique in different problem fields. It is observed through the application of the standard harmony search method that the selection of these values is problem dependent. While a certain set of values yields a good performance of the technique in one type of design problem, the same set may not present the same performance in another type of design problem. Hence it is not possible to come up with a set of values that can be used in every optimum design problem. In each problem a sensitivity analysis is required to be carried out to determine what set of values results a good performance. Adaptive harmony search method eliminates the necessity of finding the best set of parameter values. It adjusts the values of these parameters automatically during the optimization process. Before initiating the design process, a set of steel sections selected from an available profile list are collected in a design pool. Each steel section is assigned a sequence number that varies between 1 to total number of sections (N_{sec}) in the list. It is important to note that during optimization process selection of sections for design variables is carried out using these numbers. The basic components of the adaptive harmony search algorithm can now be outlined as follows.

3.1 Initialization of a Parameter Set

Harmony search method uses four parameters values of which are required to be selected by the user. This parameter set consists of a harmony memory size (hms), a harmony memory considering rate ($hmcr$), a pitch adjusting rate (par) and a maximum search number (N_{max}). Out of these four parameters, $hmcr$ and par are made dynamic parameters in adaptive harmony search method that vary from one solution vector to another. They are set to initial values of $hmcr^{(0)}$ and $par^{(0)}$ for all the solution vectors in the initial harmony memory matrix. In the standard harmony search algorithm these parameters are treated as static quantities, and they are assigned to suitable values chosen within their recommended ranges of $hmcr \in [0.70, 0.95]$ and $par \in [0.20, 0.50]$ [27-29].

3.2 Initialization of Harmony Memory Matrix

A harmony memory matrix **H** given in Eqn. (33) is randomly generated. The harmony memory matrix simply represents a design population for the solution of a problem under consideration, and incorporates a predefined number of solution vectors referred to as harmony memory size (hms). Each solution vector (harmony vector, $\mathbf{I}^i$) consists of ng design variables, and is represented in a separate row of the matrix; consequently the size of $\mathbf{H}$ is $(hms \times ng)$.

$$\mathbf{H} = \begin{bmatrix} I_1^1 & I_2^1 & \dots & I_{ng}^1 \\ I_1^2 & I_2^2 & \dots & I_{ng}^2 \\ \dots & \dots & \dots & \dots \\ I_1^{hms} & I_2^{hms} & \dots & I_{ng}^{hms} \end{bmatrix} \begin{matrix} \phi(\mathbf{I}^1) \\ \phi(\mathbf{I}^2) \\ \dots \\ \phi(\mathbf{I}^{hms}) \end{matrix} \tag{33}$$

3.3 Evaluation of Harmony Memory Matrix

The structural analysis of each solution is then performed with the set of steel sections selected for design variables, and responses of each candidate solution are obtained under the applied loads. The objective function values of the feasible solutions that satisfy all problem constraints are directly calculated from Eqn. (2). However, infeasible solutions that violate some of the problem constraints are penalized using external penalty function approach, and their objective function values are calculated according to Eqn. (34).

$$\phi = W\left[1 + \alpha\left(\sum_i g_i\right)\right] \tag{34}$$

In Eqn. (34), ϕ is the constrained objective function value, g_i is the i-th problem constraint and α is the penalty coefficient used to tune the intensity of penalization as a whole. This parameter is set to an appropriate static value of $\alpha = 1$ in the numerical examples. Finally, the solutions evaluated are sorted in the matrix in the descending order of objective function values, that is, $\phi(\mathbf{I}^1) \leq \phi(\mathbf{I}^2) \leq \ldots \leq \phi(\mathbf{I}^{hms})$.

3.4 Generating a New Harmony Vector

In harmony search algorithm the generation of a new solution (harmony) vector is controlled by two parameters ($hmcr$ and par) of the technique. The harmony memory considering rate ($hmcr$) refers to a probability value that biases the algorithm to select a value for a design variable either from harmony memory or from the entire set of discrete values used for the variable. That is to say, this parameter decides in what extent previously visited favorable solutions should be considered in comparison to exploration of new design regions while generating new solutions. At times when the variable is selected from harmony memory, it is checked whether this value should be substituted with its very lower or upper neighboring one in the discrete set. Here the goal is to encourage a more explorative search by allowing transitions to designs in the vicinity of the current solutions. This phenomenon is known as pitch-adjustment in HS, and is controlled by pitch adjusting rate parameter (par). In the standard algorithm both of these parameters are set to suitable constant values for all harmony vectors generated regardless of whether an exploitative or explorative search is indeed required at a time during the search process. On the contrary, in the adaptive algorithm a new set of values is sampled for *hmcr* and *par* parameters each time prior to improvisation (generation) of a new harmony vector, which in fact forms the basis for the algorithm to gain adaptation to varying features of the design space. Accordingly, to generate a new harmony vector in the algorithm proposed, a two-step procedure is followed consisting of (i) sampling of control parameters, and (ii) improvisation of the design vector.

3.4.1 Sampling of Control Parameters

For each harmony vector to be generated during the search process, first a new set of values are sampled for *hmcr* and *par* control parameters by applying a logistic normal distribution based variation to the average values of these parameters within the harmony memory matrix, as formulated in Eqns. (35 and 36).

$$(hmcr)^k = \left(1 + \frac{1-(hmcr)'}{(hmcr)'} . e^{-\gamma . N(0,1)}\right)^{-1} \tag{35}$$

$$(par)^k = \left(1 + \frac{1-(par)'}{(par)'} . e^{-\gamma . N(0,1)}\right)^{-1} \tag{36}$$

In Eqns. (35) and (36), $(hmcr)^k$ and $(par)^k$ represent the sampled values of the control parameters for a new harmony vector. The notation $N(0,1)$ designates a normally distributed random number having expectation 0 and standard deviation 1. The symbols $(hmcr)'$ and $(par)'$ denote the average values of control parameters within the harmony memory matrix, obtained by averaging the corresponding values of all the solution vectors within the $\mathbf{H}$ matrix, that is,

$$(hmcr)' = \frac{\sum_{i=1}^{\mu} (hmcr)^i}{(hms)} \quad , \quad (par) = \frac{\sum_{i=1}^{\mu} (par)^i}{(hms)} \tag{37}$$

Finally, the factor γ in Eqns. (35) and (36) refers to the learning rate of control parameters, which is recommended to be selected within a range of [0.25, 0.50]. In the numerical examples this parameter is set to 0.35.

In the proposed implementation, for each new vector a probabilistic sampling of control parameters is motivated around average values of these parameters $(hmcr)'$ and $(par)'$ observed in the $\mathbf{H}$ matrix. Considering the fact that the harmony memory matrix at an instant incorporates the best *hms* solutions sampled thus far during the search, the idea here is to encourage forthcoming vectors to be sampled with values that the search process has taken the most advantage in the past. The use of a logistic normal distribution provides an ideal platform in this sense because not only it guarantees the sampled values of control parameters to lie within their possible range of variation, i.e., [0, 1], but also it permits occurrence of small variations around $(hmcr)'$ and $(par)'$ more frequently than large ones. Accordingly, sampled values of control parameters mostly fall within close vicinity of the average values, yet remote values are occasionally promoted to check alternating demands of the search process.

3.4.2 Improvisation of the Design Vector

Upon sampling of a new set of values for control parameters, the new harmony vector $\mathbf{I}^k = [I_1^k, I_2^k, ..., I_{ng}^k]$ is improvised in such a way that each design variable is selected at random from either harmony memory matrix or the entire discrete set. Which one of these two sets is used for a variable is determined probabilistically in conjunction with harmony memory considering rate $(hmcr)^k$ parameter of the solution. To implement the process a uniform random number r_i is generated between 0 and 1 for each variable I_i^k. If r_i is smaller than or equal to $(hmcr)^k$, the variable is chosen from harmony memory in which case it is assigned any value from the i-th column of the $\mathbf{H}$ matrix, representing the value set of the variable in hms solutions of the matrix (Eqn. 38). Otherwise (if $r_i > (hmcr)^k$), an arbitrary value is assigned to the variable from the entire design set.

$$I_i^k = \begin{cases} I_i^k \in \{I_i^1, I_i^2, ..., I_i^{hms}\} & if\ r_i \le (hmcr)^k \\ I_i^k \in \{1,..,N_{sec}\} & if\ r_i > (hmcr)^k \end{cases} \tag{38}$$

If a design variable attains its value from harmony memory, it is checked whether this value should be pitch-adjusted or not. In pitch adjustment the value of a design variable ($I_i^{k'}$) is altered to its very upper or lower neighboring value obtained by adding ± 1 to its current value. This process is also operated probabilistically in conjunction with pitch adjusting rate $(par)^k$ parameter of the solution, Eqn. (37). If not activated by $(par)^k$, the value of the variable does not change. Pitch adjustment prevents stagnation and improves the harmony memory for diversity with a greater change of reaching the global optimum.

$$I_i^{k'} = \begin{cases} I_i^k \pm 1 & if\ r_i \le (par)^k \\ I_i^k & if\ r_i > (par)^k \end{cases} \tag{39}$$

3.5 Update of the Harmony Memory and Adaptivity

After generating the new harmony vector, its objective function value is calculated as per Eqn. (34). If this value is better (lower) than that of the worst solution in the harmony memory matrix, it is included in the matrix while the worst one is discarded out of the matrix. It follows that the solutions in the harmony memory matrix represent the best (hms) design points located thus far during the search. The harmony memory matrix is then sorted in ascending order of objective function value. Whenever a new solution is added into the harmony memory matrix, the $(hmcr)'$ and $(par)'$ parameters are recalculated using Eqn. (37). This way the harmony memory matrix is updated with the most recent information required for an efficient search and the forthcoming solution vectors are guided to make their own selection of control parameters mostly around these updated values. It should be underlined that there are no single values of control parameters that lead to the most efficient search of the algorithm throughout the process unless the design domain is completely uniform. On the contrary, the optimum values of control parameters have a tendency to change over time depending on various regions of the design space in which the search is carried out. The update of the control parameters within the harmony memory matrix enables the algorithm to catch up with the varying needs of the search process as well. Hence the most advantageous values of control parameters are adapted in the course of time automatically (i.e., by the algorithm itself), which plays the major role in the success of *adaptive* harmony search method discussed in this chapter.

3.6 Termination

The steps 3.4 and 3.5 are iterated in the same manner for each solution sampled in the process, and the algorithm terminates when a predefined number of solutions (N_{max}) is sampled.

4 Performance Evaluation of Adaptive Harmony Search Method

Performance of the adaptive harmony search algorithm presented is evaluated in the optimum design of three real size steel frames. These are 209-member industrial factory building, 568-member unbraced space steel frame and 1860-member braced space steel frame, respectively. The topology and geometry of each frame and the loadings considered in their designs are described in the relevant sections below. The design constraints in these three problems are arranged according to ASD-AISC design code specifications and the following material properties of the steel are used: modulus of elasticity (E) = 203893.6 MPa (29000 ksi) and yield stress (F_y) = 253.1 MPa (36 ksi). Each frame is designed using both standard and adaptive harmony search algorithms and the performance of the techniques is compared.

4.1 209-Member Industrial Factory Building

The first design example is an industrial factory building with 100 joints and 209 members. Shown in Figure 2 are the plan, side and 3D views of this structure. The main system of the structure consists of five identical frameworks lying 6.1 m (20 ft) apart from each other in x-z plane. Each framework consists of two side frames and a gable roof truss in between them as depicted in Figure 2 (b). The lateral stability against wind loads in x-z plane is provided with columns fixed at the base along with the rigid connections of the side frames. Hence, all the beams and columns in the side frames are designed as moment-resisting axial-flexural members.

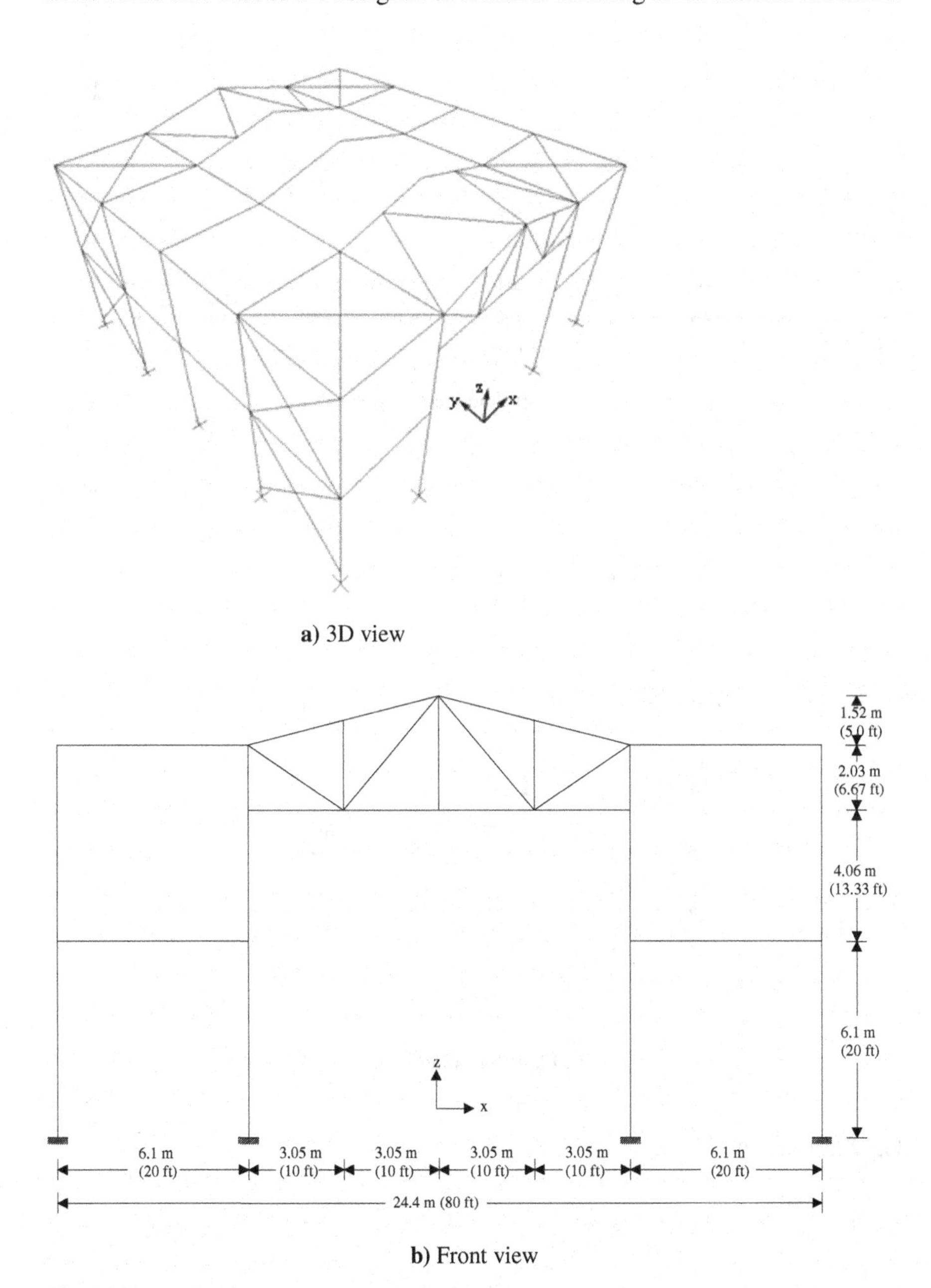

Fig. 2 209-member industrial factory building.

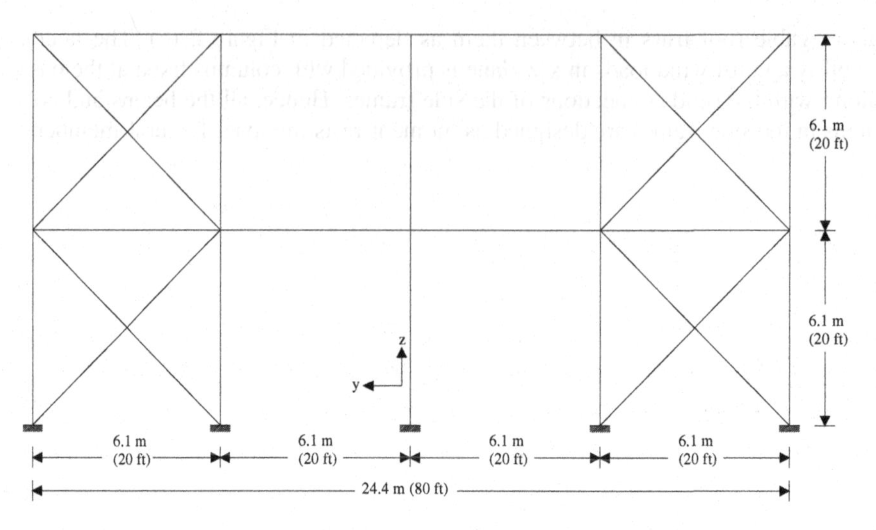

c) Side view

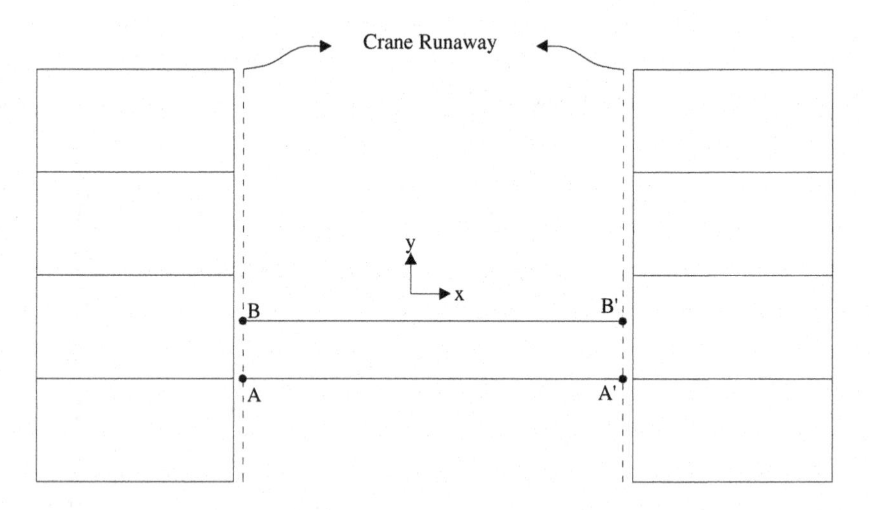

d) First floor plan view

Fig. 2 (*continued*)

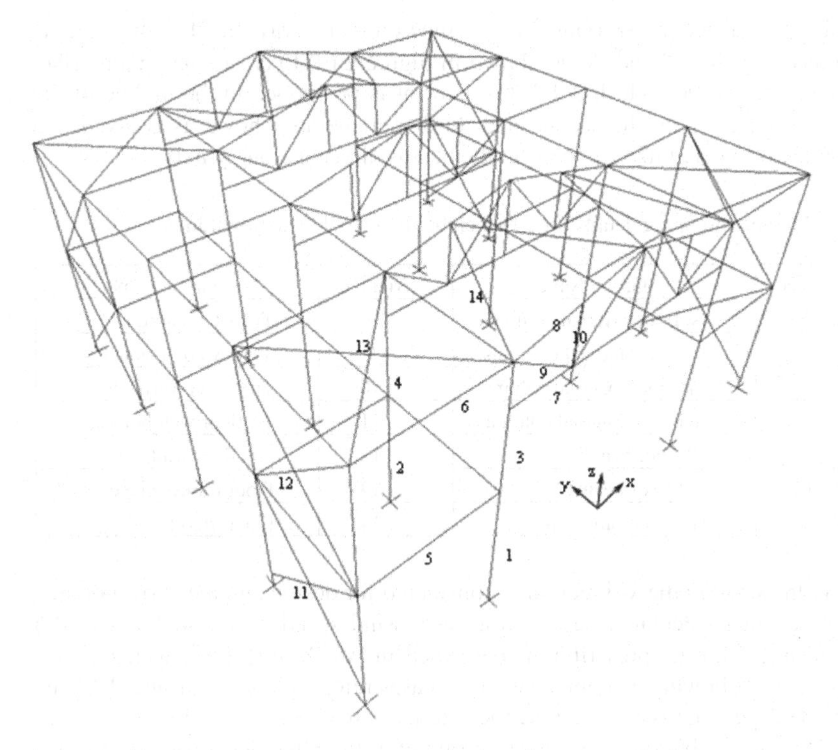

e) Member grouping

Fig. 2 *(continued)*

The gable roof truss, on the other hand, is designed to transmit only axial forces through pin-jointed connections, and hence the web and chord members in the gable roof are all designed as axial members. For longitudinal stability (along y-axis) of the structure, bracing is provided in the end bays in the walls and the roof. By employing the symmetry of the structure and fabrication requirements of structural members, the total of 209 members are collected in 14 member groups (independent size variables). The member grouping details are presented in Table 1 and Figure 2 (e).

Three different types of loads are considered for the design of the industrial building; namely dead, crane and wind loads. A design dead load of 1.2 kN/m^2 is assumed to be acting on both floors of the side frames, resulting in uniformly distributed loads of 14.63 kN/m (1004.55 lb/ft) and 7.32 kN/m (502.27 lb/ft) on the interior and exterior beams of the side frames. The dead weights of the gable roofs are neglected due to relatively light weight of these components. The crane load is modeled as two pairs of moving live loads acting on both sides of the crane runway beams as shown in Figure 2 (d). Each pair consists of a concentrated load of 280 kN (62.9 kip) and a couple moment of 75 kN.m (5532 kip.ft). In the study the crane load is represented in two distinct load cases referred to as CL1 and CL2 by choosing two different positions for the crane on its runway. In CL1, the crane is positioned at points A and A′ as shown in Figure 2 (d) to create maximum effect on the second framework. In CL2, however, it is positioned in the middle of the runway beam between the second and third frameworks (shown as B and B′ in Figure 2 (d)) to maximize response in the beams directed along y-axis.

Table 1 Member grouping details for 209-member industrial factory building.

Member	Group Name	Member	Group name
1	1st floor external columns	8	Truss top chord
2	1st floor internal columns	9	Truss web diagonals
3	2nd floor external columns	10	Truss web verticals
4	2nd floor internal columns	11	1st floor wall braces
5	1st floor beams	12	2nd floor wall braces
6	2nd floor beams	13	Floor frames braces
7	Truss bottom chord	14	Floor truss braces

Only the wind in the x-direction is considered for design and the corresponding wind forces are calculated based on a basic wind speed of $V = 46.94$ m/s (105 mph) in line with the prescriptions described in ASCE 7-05 [30], which is discussed in the following example. Two load cases referred to as WL1 and WL2 are generated depending on the sign of the internal wind pressure exerted on the external faces of the building, as shown in Figure 3. In both cases, it is assumed that wind causes a positive compression pressure on windward face, while it causes a negative suction effect on leeward face as well as on side walls of the building. In WL1 the suction effect is considered for the entire roof surface, whereas in WL2 one part of the roof is subjected to compression pressure. From amongst the five load cases (DL, CL1, CL2, WL1 and WL2), a total of six load combinations are generated for the strength design of each structural member in accordance with ASD-AISC [11] specification, as follows:

(i) 1.0DL + 1.0CL1
(ii) 1.0DL + 1.0CL1 + 1.0WL1
(iii) 1.0DL + 1.0CL1 + 1.0WL2
(iv) 1.0DL + 1.0CL2
(v) 1.0DL + 1.0CL2 + 1.0WL1
(vi) 1.0DL + 1.0CL2 + 1.0WL2

All members are sized using the standard sections in AISC. Accordingly, the beam and column members are selected from wide-flange sections (W), and side wall and roof bracings are selected from back to back equal leg double angle sections. The combined stress, stability and geometric constraints are imposed according to the provisions of ASD-AISC [11]. In addition, displacements of all the joints in x and y directions are limited to 3.43 cm (1.25 in), and the upper limit of inter-story drifts is set to 1.52 cm (0.6 in).

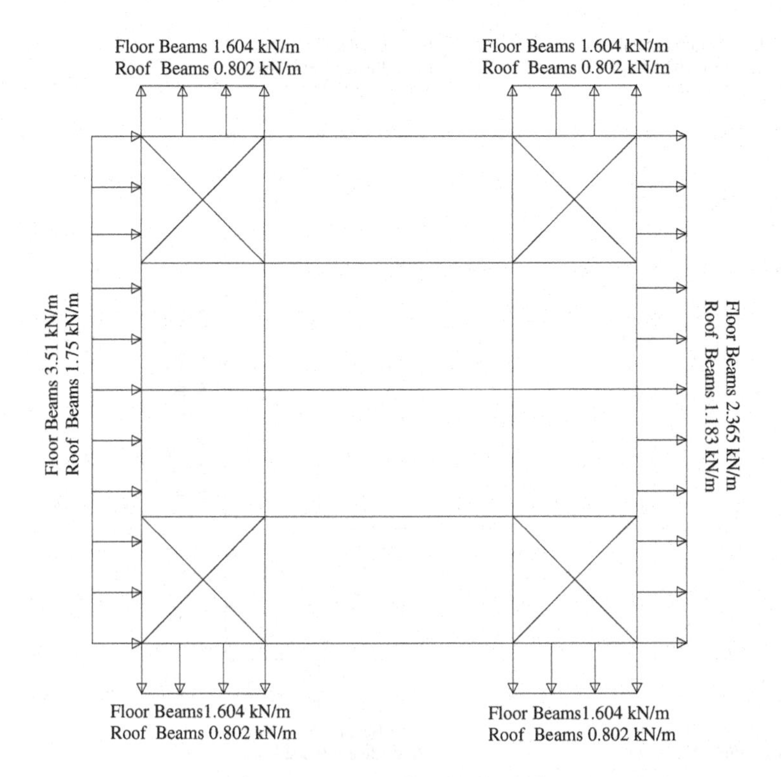

i- wind loads on floor and roof beams (case 1)

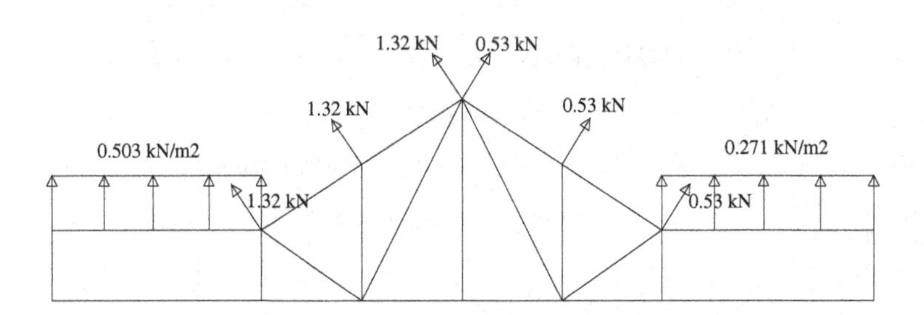

ii- wind loads on the roof (case 1)

a) The first wind load case (WL1)

Fig. 3 The two wind load cases considered for the design of 209-member industrial factory building.

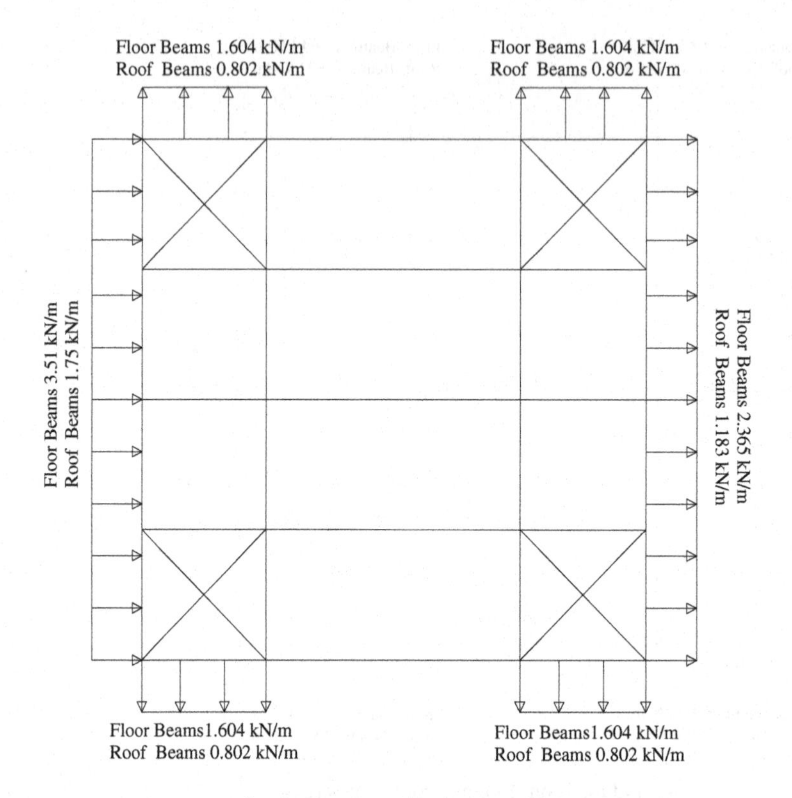

i- wind loads on floor and roof beams (case 2)

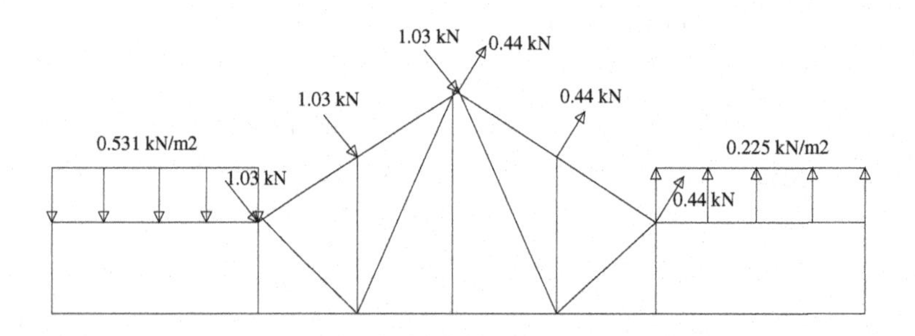

ii- wind loads on the roof (case 2)

b) The second wind load case (WL2)

Fig. 3 (*continued*)

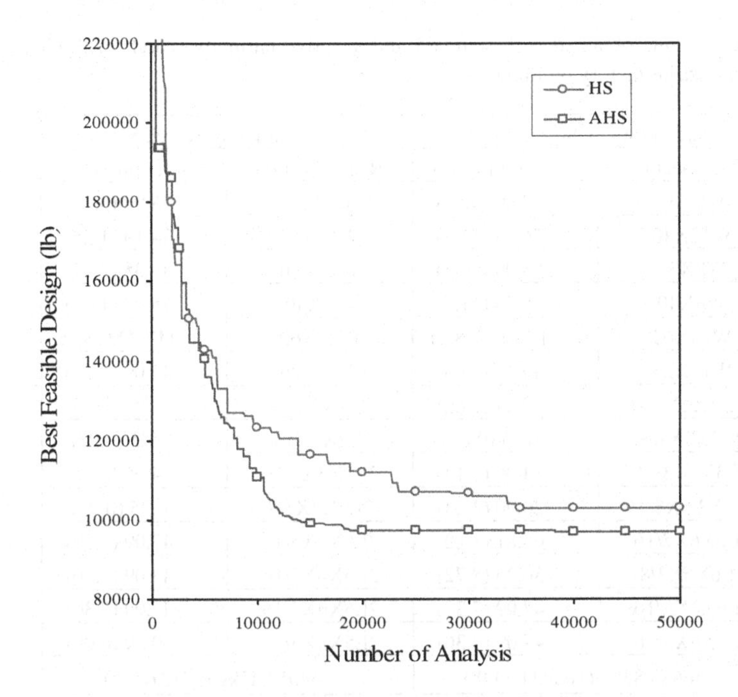

Fig. 4 The design history for 209-member industrial factory building.

The industrial factory building described above is designed by using standard and adaptive harmony search techniques. The harmony memory size *hms*, harmony memory considering rate *hmcr* and pitch adjustment rate *par* is taken as 50, 0.90 and 0.30 in the standard harmony search method while these parameters are adjusted dynamically in the adaptive harmony search technique. The maximum number of iterations is taken as 50000 in both algorithms in order to provide equal opportunity for both algorithms for attaining the global optimum. The design history of both runs is shown in Figure 4. It is apparent from the figure that adaptive harmony method exhibits a better convergence rate and obtains lighter frame. The minimum weight of the steel frame is attained as 46685.83kg by the standard harmony search method while the same weight is obtained as 44053.45kg by the adaptive harmony search algorithm which is 5.6 % lighter. It is also apparent from the figure that adaptive harmony search method approaches to the vicinity of the minimum weight in early iterations of the optimum design process while the standard harmony search method reduces the frame weight in a steady manner until the end of the design process. The steel section designations determined by the both methods for each member groups of the frame are given in Table 2.

Table 2 Optimum designs obtained with standard and adaptive harmony search methods for 209-member industrial factory building.

Group Number	Standard Harmony Search Method		Adaptive Harmony Search Method	
	Ready Section	Area, (cm^2) (in^2)	Ready Section	Area, (cm^2) (in^2)
1	W8X31	58.90 (9.13)	W8X31	58.90 (9.13)
2	W12X40	76.13 (11.8)	W10X39	74.19 (11.5)
3	W8X31	58.90 (9.13)	W12X26	49.35 (7.65)
4	W8X40	75.48 (11.7)	W8X40	75.48 (11.7)
5	W24X62	117.42 (18.2)	W24X62	117.42 (18.2)
6	W12X26	49.35 (7.65)	W10X26	49.09 (7.61)
7	2L2.5X2X3/16	10.44 (1.62)	2L2X2X1/8	6.25 (0.97)
8	2L2X2X1/8	6.25 (0.97)	2L2X2X1/8	6.25 (0.97)
9	2L3X3X3/16	14.06 (2.18)	2L3X3X3/16	14.06 (2.18)
10	2L3X2.5X5/16	20.90 (3.24)	2L2X2X1/8	6.25 (0.97)
11	2L6X6X7/16	65.81 (10.2)	2L6X6X5/16	47.09 (7.30)
12	2L6X6X3/8	56.26 (8.72)	2L6X6X5/16	47.09 (7.30)
13	2L6X6X5/16	47.09 (7.30)	2L6X6X5/16	47.09 (7.30)
14	2L6X6X5/16	47.09 (7.30)	2L5X5X5/16	39.09 (6.06)
Weight	46685.83kg (102924.73 lb)		44053.45kg (97121.3 lb)	

4.2 568-Member Unbraced Space Steel Frame

The second design example shown in Figures 5 (a-d) is a 10-story unbraced space steel frame consisting of 256 joints and 568 members. This problem has been formerly studied in Hasançebi et al. [25] to evaluate the performance of various meta-heuristic search techniques in real size optimum design of steel frameworks. The objective in this problem is then to compare the performance of adaptive harmony search method with those of other meta-heuristic search techniques.

The columns in a story are collected in three member groups as corner columns, inner columns and outer columns, whereas beams are divided into two groups as inner beams and outer beams. The corner columns are grouped together as having the same section in the first three stories and then over two adjacent stories thereafter, as are inner columns, outer columns, inner beams and outer beams. This results in a total of 25 distinct member groups as shown in Figure 5 (d). The columns are selected from the complete W-shape profile list consisting of 297 ready sections, whereas a discrete set of 171 economical sections selected from W-shape profile list based on area and inertia properties is used to size beam members.

The frame is subjected to various gravity loads in addition to lateral wind forces. The gravity loads acting on floor slabs cover dead (DL), live (LL) and snow (SL) loads, which are applied as uniformly distributed loads on the beams using load distribution formulas developed for slabs. All the floors, except the roof, are subjected to a design dead load of 2.88 kN/m^2 (60.13 lb/ft^2) and a design live load of 2.39 kN/m^2 (50 lb/ft^2). The beams of the roof level are subjected to the design dead load plus snow load. The design snow load is computed using the following equation in ASCE 7-05 [30]:

$$p_s = 0.7C_s C_e C_t I p_g \tag{40}$$

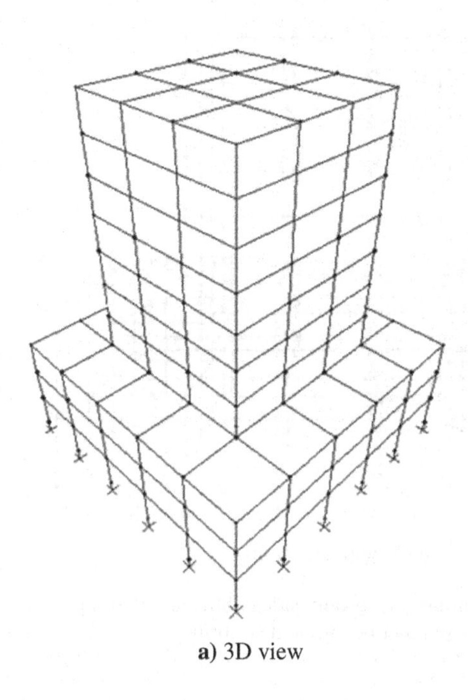

a) 3D view

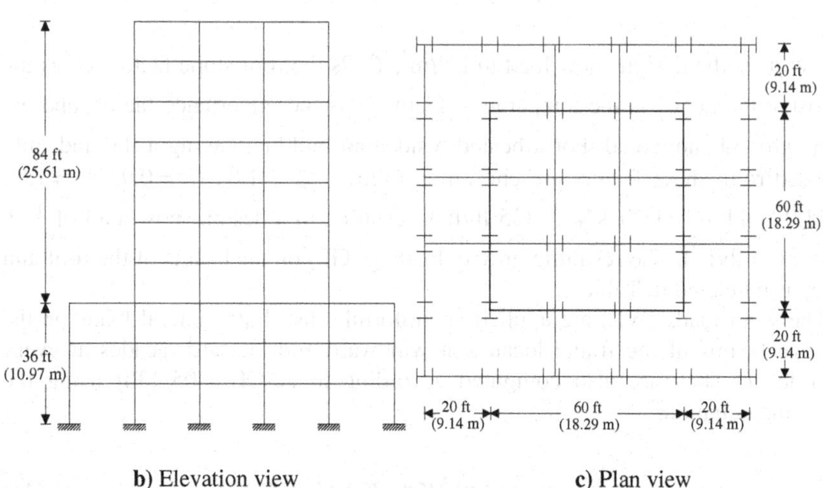

b) Elevation view c) Plan view

Fig. 5 568-member unbraced space steel frame.

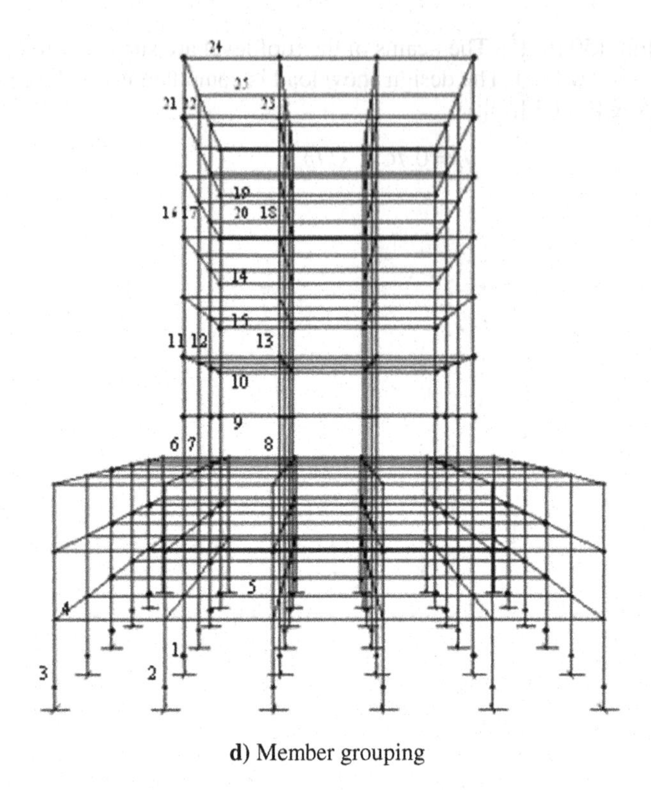

d) Member grouping

* 1st group: inner columns, 2nd group: side columns, 3rd group: corner columns 4th group: outer beams 5th group: inner beams, and so forth.

Fig. 5 (*continued*)

where p_s is the design snow load in kN/m^2, C_s is the roof slope factor, C_e is the exposure factor, C_t is the temperature factor, I is the importance factor, and p_g is the ground snow load. For a heated residential building having a flat and fully exposed roof, these factors are chosen as follows: $C_s = 1.0$, $C_e = 0.9$, $C_t = 1.0$, $I = 1.0$, and $p_g = 1.20$ kN/m^2 (25 lb/ft^2), resulting in a design snow load of 1.20 kN/m^2 (25 lb/ft^2). The resulting gravity loading (GL) on the beams of the roof and floors is tabulated in Table 3.

The wind loads (WL) are applied as uniformly distributed lateral loads on the external beams of the frame located at windward and leeward facades at every floor level. They are also computed according to ASCE 7-05 [30] using the following equation:

$$p_w = (0.613 K_z K_{zt} K_d V^2 I)(GC_p) \tag{41}$$

Table 3 Gravity loading on the beams of 568-member unbraced space steel frame.

Beam Type	Uniformly Distributed Load	
	Outer Beams	Inner Beams
	kN/m (lb/ft)	kN/m (lb/ft)
Roof beams	7.38 (505.879)	14.77 (1011.74)
Floor beams	10.72 (734.20)	21.44 (1468.40)

Table 4 Wind loading on 568-member unbraced space steel frame.

Floor No	Windward	Leeward
	kN/m (lb/ft)	kN/m (lb/ft)
1	1.64 (112.51)	1.86 (127.38)
2	1.88 (128.68)	1.86 (127.38)
3	2.10 (144.68)	1.86 (127.38)
4	2.29 (156.86)	1.86 (127.38)
5	2.44 (167.19)	1.86 (127.38)
6	2.57 (176.13)	1.86 (127.38)
7	2.69 (184.06)	1.86 (127.38)
8	2.79 (191.21)	1.86 (127.38)
9	2.89 (197.76)	1.86 (127.38)
10	1.49 (101.90)	0.93 (63.69)

where p_w is the design wind pressure in kN/m^2, K_z is the velocity exposure coefficient, K_{zt} is the topographic factor, K_d is the wind direction factor, V is the basic wind speed, G is the gust factor, and C_p is the external pressure coefficient. Assuming that the building is located in a flat terrain with a basic wind speed of $V = 46.94$ m/s (105 mph) and exposure category B, the following values are used for these parameters: $K_{zt} = 1.0$, $K_d = 0.85$, $I = 1.0$, $G = 0.85$, and $C_p = 0.8$ for windward face and -0.5 for leeward face. The calculated wind loads at every floor level are presented in Table 4.

The gravity and wind forces are combined under two loading conditions. In the first loading condition, the gravity loading is applied with the wind loading acting along x-axis (1.0GL + 1.0WL-x), whereas in the second one wind loading is acted along y-axis (1.0GL + 1.0WL-y). The combined stress, stability, displacement and geometric constraints are imposed according to the provisions of ASD-AISC [11].

The optimum design of the unbraced space steel frame described above is carried out using the adaptive harmony search algorithm as well as six different meta-heuristic techniques. These meta-heuristic techniques are evolutionary strategies (ES), tabu search optimization (TSO), simulated annealing (SA), ant colony optimization (ACO), simple genetic algorithm (SGA) and particle swarm optimizer (PSO). In each optimization technique the number of iterations is taken as 50000 in order to allow equal opportunity to every technique to grasp the global

Table 5 Comparison of optimum designs obtained by various meta-heuristic search techniques for 568-member unbraced space steel frame.

Member groups	Ready Sections in designs obtained with each metaheuristic optimization technique						
	ESs	AHS	TSO	SA	ACO	SGA	PSO
1	W14X193	W14X176	W14X193	W14X193	W14X193	W14X193	W14X159
2	W8X48	W14X48	W8X48	W8X48	W8X48	W8X48	W24X76
3	W10X39	W10X39	W8X40	W8X40	W10X45	W10X39	W10X39
4	W10X22	W10X22	W10X22	W10X22	W10X22	W10X26	W10X22
5	W21X50	W24X55	W21X50	W21X44	W21X50	W21X50	W24X55
6	W10X54	W12X65	W10X54	W12X65	W14X61	W18X76	W12X72
7	W14X109	W14X145	W14X120	W14X145	W14X120	W14X109	W27X146
8	W14X176	W14X159	W14X159	W14X145	W40X192	W40X192	W27X217
9	W18X40	W14X30	W21X44	W24X68	W18X35	W18X40	W18X40
10	W18X40	W18X40	W18X40	W24X55	W18X40	W21X50	W18X40
11	W10X49	W10X54	W10X45	W10X49	W12X58	W12X65	W18X71
12	W14X90	W14X90	W14X90	W14X90	W12X96	W21X111	W21X101

Table 5 *(continued)*

13	W14X109	W14X120	W12X120	W14X120	W12X136	W12X152	W14X176
14	W14X30	W14X34	W21X44	W16X36	W12X30	W12X30	W14X34
15	W16X36	W18X40	W16X36	W16X40	W21X44	W16X40	W21X44
16	W12X45	W8X31	W10X33	W12X40	W8X58	W14X68	W12X65
17	W12X65	W12X65	W12X65	W12X65	W18X76	W18X76	W10X68
18	W10X22	W18X35	W14X34	W12X26	W12X35	W8X28	W12X35
19	W12X79	W12X79	W12X79	W12X72	W10X88	W10X88	W12X79
20	W14X30	W14X30	W14X30	W16X36	W14X30	W16X36	W14X38
21	W8X35	W10X22	W10X39	W8X24	W8X58	W8X48	W10X39
22	W10X39	W10X45	W12X45	W10X49	W8X40	W14X34	W8X31
23	W8X31	W8X31	W12X35	W8X24	W8X31	W12X30	W12X96
24	W8X18	W10X22	W6X20	W12X26	W8X24	W8X21	W12X26
25	W14X30	W12X26	W12X26	W12X26	W16X45	W18X35	W12X26
Weight, kg (lb)	228588.33 (503953.63)	232301.20 (512139.16)	235167.52 (518458.35)	238756.51 (526370.76)	241470.31 (532353.70)	245564.80 (541380.54)	253260.23 (558346.15)

optimum. The design history of each run by each technique is shown in Figure 6 and the minimum weights as well as W-section designations obtained for each members group is given in Table 5. Inspection of the minimum weights reveals the fact that the lightest frame is attained by the evolutionary strategies and the optimum result obtained by the adaptive harmony search algorithm is the second best among all the meta-heuristic algorithms considered in this study. This clearly indicates that the enhancements carried out in the standard harmony search method have certainly improved the performance of the technique. In fact the optimum design attained by the standard harmony search method for the same frame was 259072.31 kg (571159.66 lb) as given in [25] which was the heaviest among all. The minimum weight found in this study is only 1.6% heavier than the one obtained by evolutionary strategies algorithm.

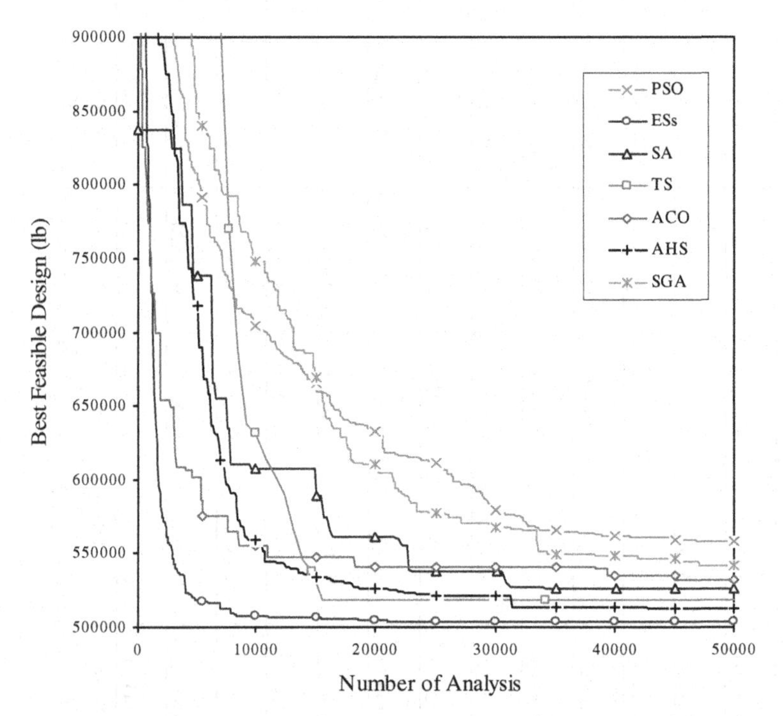

Fig. 6 The design history for meta-heuristic search algorithms used in the optimum design of 568-member unbraced space steel frame.

4.3 1860-Member Braced Space Steel Frame

The last design example considered in this section is 36-story braced space steel frame consisting of 814 joints and 1860 members. The side, plan and 3D views of the frame as well as member grouping details are shown in Figures 7 (a-d). An economical and effective stiffening of the frame against lateral forces is achieved through exterior diagonal bracing members located on the perimeter of the building, which also participate in transmitting the gravity forces.

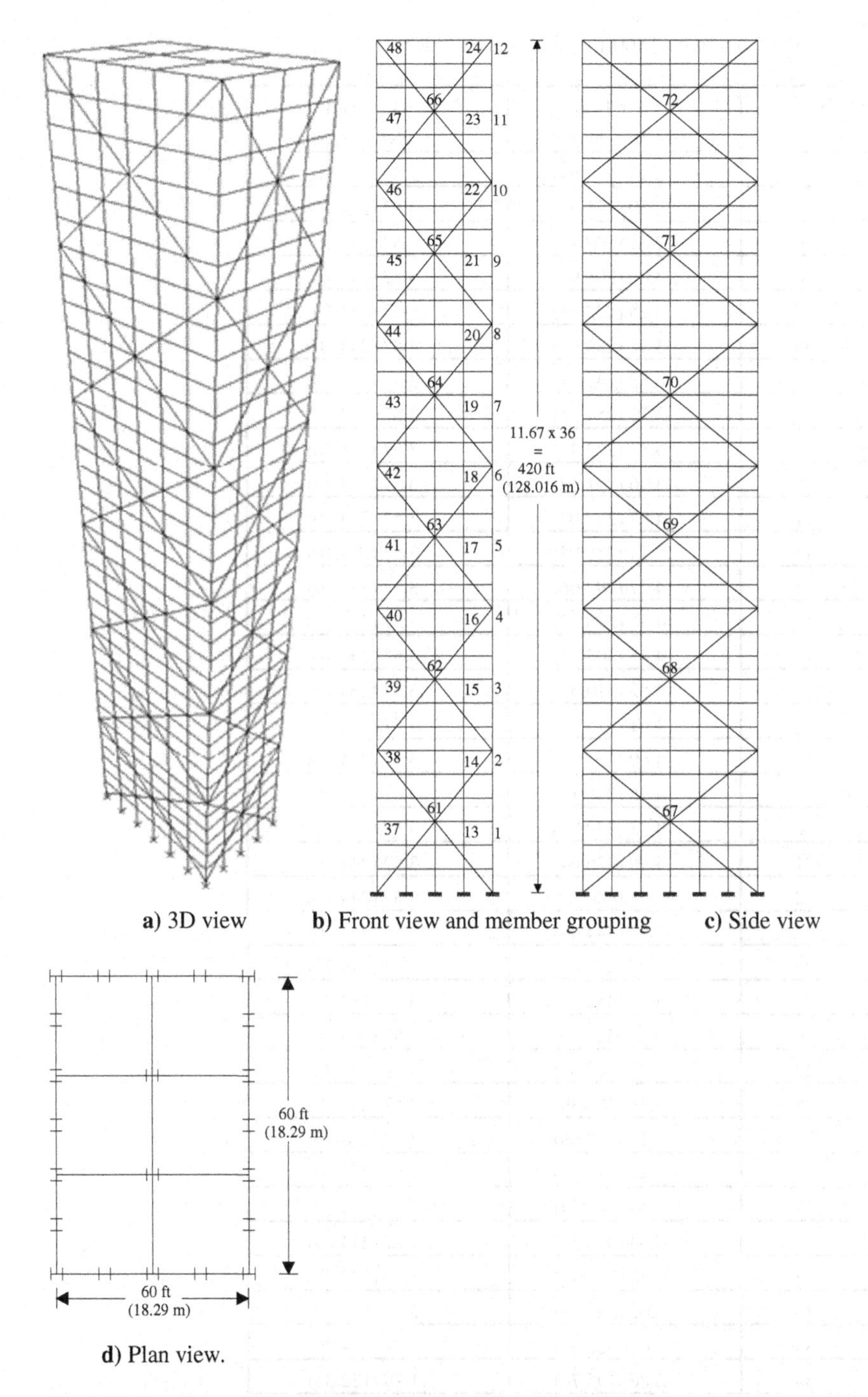

Fig. 7 1860-member braced space steel frame.

Table 6 Wind loading on 1860-member braced space steel frame.

Floor	Windward kN/m (lb/ft)	Leeward kN/m (lb/ft)
1	2.05 (140.64)	3.57 (244.70)
2	2.50 (171.44)	3.57 (244.70)
3	2.81 (192.49)	3.57 (244.70)
4	3.05 (208.98)	3.57 (244.70)
5	3.25 (222.74)	3.57 (244.70)
6	3.42 (234.65)	3.57 (244.70)
7	3.58 (245.22)	3.57 (244.70)
8	3.72 (254.75)	3.57 (244.70)
9	3.85 (263.47)	3.57 (244.70)
10	3.96 (271.52)	3.57 (244.70)
11	4.07 (279.02)	3.57 (244.70)
12	4.18 (286.04)	3.57 (244.70)
13	4.27 (292.66)	3.57 (244.70)
14	4.36 (298.92)	3.57 (244.70)
15	4.45 (304.87)	3.57 (244.70)
16	4.53 (310.55)	3.57 (244.70)
17	4.61 (315.97)	3.57 (244.70)
18	4.69 (321.18)	3.57 (244.70)
19	4.76 (326.18)	3.57 (244.70)
20	4.83 (330.99)	3.57 (244.70)
21	4.90 (335.64)	3.57 (244.70)
22	4.97 (340.13)	3.57 (244.70)
23	5.03 (344.48)	3.57 (244.70)
24	5.09 (348.69)	3.57 (244.70)
25	5.15 (352.78)	3.57 (244.70)
26	5.21 (356.76)	3.57 (244.70)
27	5.27 (360.62)	3.57 (244.70)
28	5.32 (364.39)	3.57 (244.70)
29	5.37 (368.06)	3.57 (244.70)
30	5.43 (371.65)	3.57 (244.70)
31	5.48 (375.14)	3.57 (244.70)
32	5.53 (378.56)	3.57 (244.70)
33	5.58 (381.90)	3.57 (244.70)
34	5.62 (385.18)	3.57 (244.70)
35	5.67 (388.38)	3.57 (244.70)
36	2.86 (195.76)	1.79 (122.35)

Table 7 Gravity loading on the beams of 1860-member braced steel space frame.

Beam Type	Uniformly Distributed Load, kN/m (lb/ft)		
	Dead Load	Live Load	Snow Load
Roof beams	22.44 (1536.66)	N.A	5.88 (402.50)
Floor beams	22.44 (1536.66)	18.66 (1277.78)	N.A

The wide-flange (W) profile list consisting of 297 ready sections is used to size column members, while beams and diagonals are selected from discrete sets of 171 and 147 economical sections selected from wide-flange profile list based on area and inertia properties in the former, and on area and radii of gyration properties in the latter. The 1860 frame members are collected in 72 different member groups, considering the symmetry of the structure and practical fabrication requirements. That is, the columns in a story are collected in three member groups as corner columns, inner columns and outer columns, whereas beams are divided into two groups as inner beams and outer beams. The corner columns are grouped together as having the same section over three adjacent stories, as are inner columns, outer columns, inner beams and outer beams. Bracing members on each facade are designed as three-story deep members, and two bracing groups are specified in every six stories.

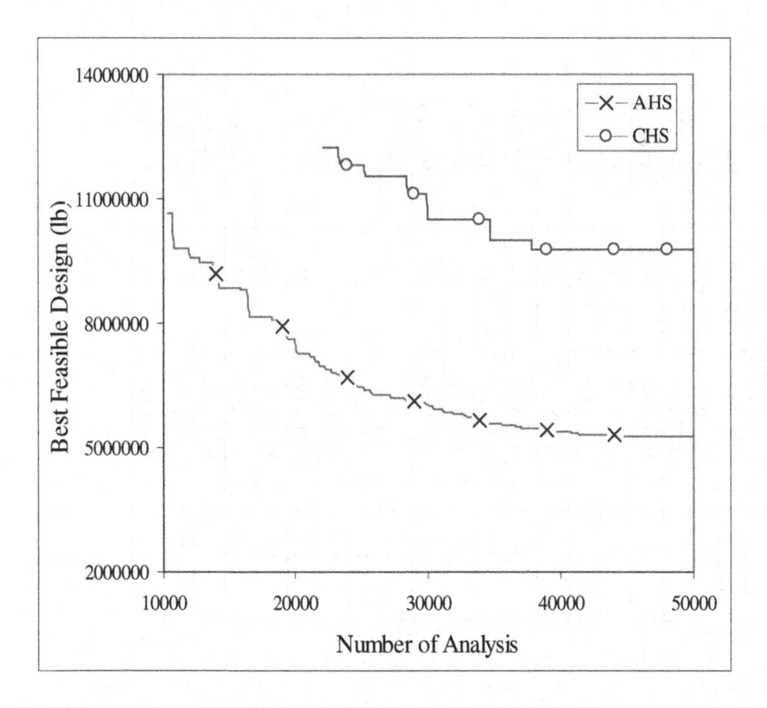

Fig. 8 The design history for standard and adaptive harmony search methods used in the optimum design of 1860-member braced space steel frame.

Table 8 Comparison of optimum designs obtained with classical and adaptive harmony search methods for 1860-member braced steel space frame.

Member Group	Standard Harmony Search Method			Member Group	Adaptive Harmony Search Method		
	Ready Section	Member Group	Ready Section		Ready Section	Member Group	Ready Section
1	W27X258	37	W18X35	1	W24X370	37	W12X14
2	W40X268	38	W30X148	2	W12X336	38	W16X26
3	W24X450	39	W18X35	3	W12X305	39	W14X22
4	W24X250	40	W36X210	4	W12X190	40	W14X26
5	W27X146	41	W10X30	5	W14X193	41	W27X102
6	W36X848	42	W36X160	6	W24X117	42	W16X31
7	W33X201	43	W14X30	7	W12X79	43	W18X35
8	W33X201	44	W33X141	8	W12X79	44	W18X35
9	W14X455	45	W12X16	9	W40X244	45	W40X167
10	W36X359	46	W12X26	10	W18X86	46	W14X22
11	W33X201	47	W33X152	11	W12X96	47	W24X68
12	W24X176	48	W21X50	12	W8X28	48	W16X26
13	W30X581	49	W21X57	13	W33X424	49	W24X62
14	W36X798	50	W33X387	14	W40X436	50	W27X94
15	W14X665	51	W40X167	15	W40X324	51	W24X68
16	W36X798	52	W30X235	16	W36X280	52	W24X76
17	W36X393	53	W30X90	17	W33X318	53	W24X76
18	W36X848	54	W27X84	18	W33X291	54	W30X116

Table 8 (*continued*)

19	W36X848	55	W33X141	19	W40X277	55	W27X94
20	W36X848	56	W36X527	20	W24X250	56	W21X83
21	W33X424	57	W40X167	21	W36X260	57	W30X90
22	W36X848	58	W33X152	22	W33X291	58	W44X198
23	W36X848	59	W44X248	23	W27X235	59	W44X285
24	W40X480	60	W30X124	24	W12X170	60	W24X68
25	W36X848	61	W14X605	25	W14X665	61	W14X455
26	W36X798	62	W14X730	26	W36X798	62	W14X398
27	W36X527	63	W36X328	27	W36X720	63	W40X328
28	W36X848	64	W30X173	28	W33X619	64	W14X233
29	W14X500	65	W14X176	29	W40X531	65	W14X109
30	W27X281	66	W21X166	30	W36X439	66	W12X72
31	W36X798	67	W14X311	31	W27X494	67	W40X328
32	W36X848	68	W33X387	32	W33X619	68	W14X283
33	W36X848	69	W36X300	33	W21X364	69	W14X233
34	W40X324	70	W40X249	34	W40X297	70	W40X192
35	W36X527	71	W40X249	35	W36X245	71	W40X192
36	W36X798	72	W30X261	36	W14X283	72	W14X132
Weight	4438172.37 kg (9784496.01 lb)			2383604.61 kg (5254949.08 lb)			

The 1860-member braced space steel frame is subjected to two loading conditions of combined gravity and wind forces. These forces are computed as per ASCE 7-05 based on the following design values: a design dead load of 2.88 kN/m^2 (60.13 lb/ft^2), a design live load of 2.39 kN/m^2 (50 lb/ft^2), a ground snow load of 1.20 kN/m^2 (25 lb/ft^2) and a basic wind speed of 55.21 m/s (123.5mph). Lateral (wind) loads acting at each floor level on windward and leeward faces of the frame are tabulated in Table 6 and the gravity loading on the beams of roof and floors is given in Table 7. In the first loading condition, gravity loads are applied together with wind loads acting along x-axis (1.0 GL + 1.0WL-x), whereas in the second one they are applied with wind loads acting along y-axis (1.0 GL + 1.0WL-y). The combined stress, stability and geometric constraints are imposed according to the provisions of ASD-AISC. The joint displacements in x and y direction are restricted to 32.0 cm (12.6 in) which is obtained as height of frame/400. Furthermore, story drift constraints are applied to each story of the frame which is equal to height of each story/400.

The 1860-member braced space steel frame is designed separately by using both standard and adaptive harmony search method. In the standard harmony search method the harmony memory size, harmony memory considering rate and pitch adjustment rate are taken as 50, 0.90 and 0.10 respectively. The maximum number of iteration is 50000. The design history of both runs is shown in Figure 8 and the optimum designs obtained by the both algorithm is given in Table 8. The minimum weight for the frame is determined as 2383604.61 kg by the adaptive harmony search method while standard harmony search algorithm arrived at 4438172.37 kg which is 46.3% heavier. It is apparent that in optimum design problems where the number of design variables relatively large, standard harmony search method do not perform well and adaptive harmony search technique dissipates this drawback. Figure 8 clearly demonstrates the better performance of the adaptive harmony search method and verifies the above fact.

5 Comparison of Code Based Optimum Designs

Figure 9 shows plan and elevation views of a 85-member moment resisting planar steel frame, which actually represents one of the interior frameworks of a steel building along the short side. The 85 members are grouped into total of 21 independent size variables to satisfy practical fabrication requirements, such that the exterior columns are grouped together as having the same section over two adjacent stories, as are interior columns and beams, as indicated in Figure 9.

The frame is only subjected to gravity loads, which are computed as per ASCE 7-05 [30] based on the following design values: a design dead load of 2.88 kN/m^2 (60.13 lb/ft^2), a design live load of 2.39 kN/m^2 (50 lb/ft^2) and a ground snow load of 1.20 kN/m^2 (25 lb/ft^2). The unfactored distributed gravity loads on the beams of the roof and floors are tabulated in Table 9. The load and combination factors are applied according to each code specification used to size the frame members, as follows: 1.0DL + 1.0L + 1.0SL for ASD-AISC; 1.2DL + 1.6LL + 0.5SL for LRFD-AISC; and 1.4DL + 1.6LL + 1.6SL for BS5950.

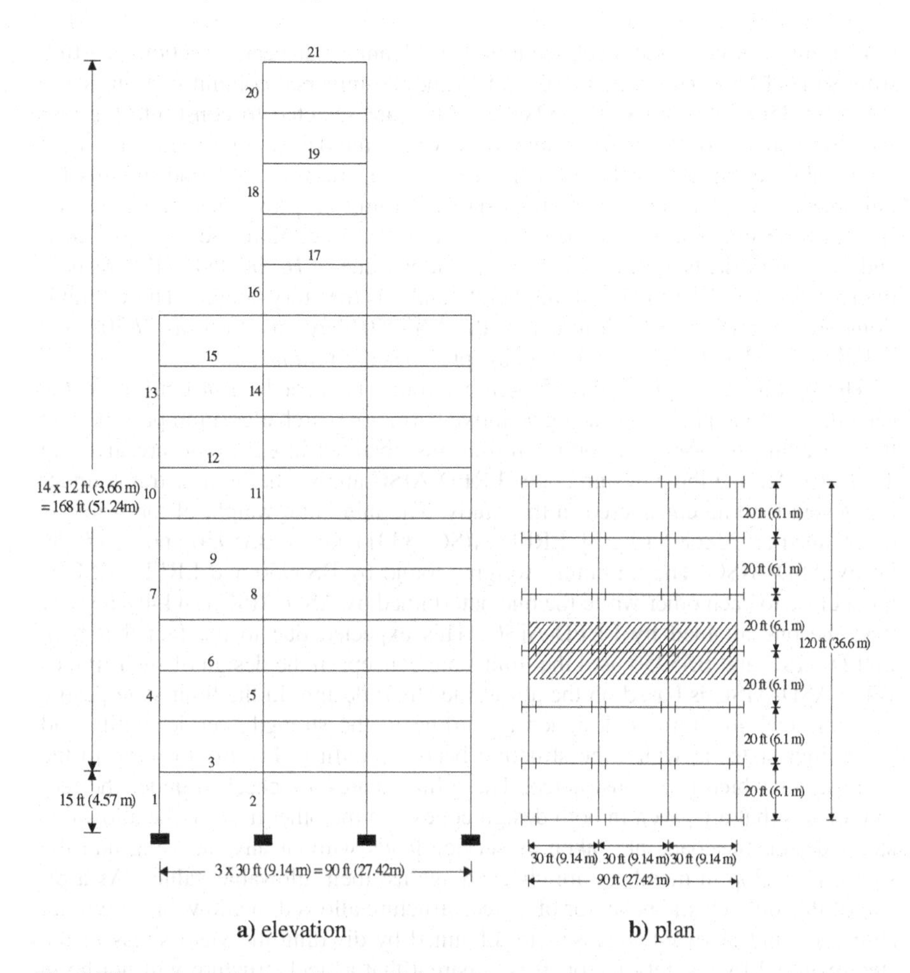

Fig. 9 85-member unbraced planar steel frame.

Table 9 Gravity loading on the beams of 85-member unbraced planar steel frame.

Beam Type	Uniformly Distributed Load, kN/m (lb/ft)		
	Dead Load	Live Load	Snow Load
Roof beams	7.47 (512.22)	N.A.	1.96 (134.17)
Floor beams	7.47 (512.22)	6.21 (425.93)	N.A.

In the ASD-AISC and LRFD-AISC code applications the wide-flange (W) profile list consisting of 297 ready sections is used to size column members, while beams are selected from discrete sets of 171 economical sections selected from wide-flange profile list based on area and inertia properties in the optimum design of the frame. In the case of British Code it is common practice to use universal beam (UB) sections for beams and universal column (UC) sections for columns of steel frames. Among the steel sections list 64 universal beam sections starting from 914x419x388UB to 254x102x28UB and 32 universal column sections starting from 356x406x634UC to 152x152x23UC are selected to constitute the discrete set consists of 96 steel sections from which the design algorithm selects the sectional designations for the frame members. The stress, strength and stability requirements of all members are imposed according to provisions of each code specification employed, as outlined in Section 2. In addition, the top story drift and inter-story drifts are limited to a maximum value of $H/400$ and $h/400$, respectively, where H is the total building height and h is the story height. These limitations are somewhat little different in the BS5950 they are given as $H/300$ and $h/300$ where H and h are the same as the previous definitions.

The optimum design of the 85-member frame is carried out according to the each design code provisions using adaptive harmony search algorithm presented in the preceding sections. The optimum designs obtained in each case are given in Table 10. Among three design codes LRFD-AISC attains the lightest frame under the design loading considered in this study. The minimum weight of the frame is determined as 32868.54 kg by LRFD-AISC, 33011 kg by BS5950 and 47472.66 kg by ASD-AISC. The minimum weights found by BS5950 and LRFD-AISC is quite close to each other while the one determined by ASD-AISC is 44.4% heavier than the one attained by LRFD-AISC. This expected due to the fact that both LRFD-AISC and BS5950 uses the limit state concept in the design of steel frames while ASD-AISC is based on the allowable stress design. In the limit state design concept steel structure is designed according to the strength, serviceability and other limit states at which the structure becomes unfit to be able to serve to the purpose for which it is constructed. These limit states are checked under the factored loads that are given in both design codes. On the other hand in the allowable stress design the loads are taken as service loads without any factoring and the stresses develop in members are checked against their allowable values. As a result of this only elastic behavior of a steel structure allowed in allowable stress design code and allowable stresses are obtained by dividing the yield stress of the steel material by a safety factor. It is apparent that a steel structure will not be in an unsafe condition even though stresses in some of its members exceed the allowable stress values because of the fact that allowable stresses are much lower than the yield stress of the steel material. On the other hand, in the limit state design concepts the service loads are increased by load factors and the stresses develop under these loads are allowed to reach to yield strength values of steel material. Consequently, the design based on the limit state design concepts yields a lighter structure due to the fact that it takes into account the realistic behavior of steel structures. It is worthwhile to state the fact that in this study only gravity loadings are considered, the other loading cases are not considered in the optimum design. The difference between the optimum designs obtained according to ASD and LRFD design codes may be less when all the loading cases are considered. However, it is known that design codes based on the limit state concepts results in lighter designs [31]. It is interesting to notice that the optimum design obtained by considering the design constraints from LRFD-AISC design code is less but not very much different than the one obtained considering the design constraints from

Table 10 The comparison of optimum designs produced according to ASD-AISC, LRFD- AISC and BS5950 design codes for 85-member unbraced planar steel frame.

Member group	BS5950		ASD-AISC		LRFD-AISC	
	Section Designation	Area, cm^2 (in^2)	Section Des-	Area, cm^2 (in^2)	Section Des-	Area, cm^2 (in^2)
1	457X152X52 UB	66.5 (10.30)	W12X72	136.13 (21.1)	W16X31	58.90 (9.13)
2	457X152X52 UB	66.5 (10.30)	W14X68	129.03 (20.0)	W16X40	75.80 (11.75)
3	457X152X52 UB	66.5 (10.30)	W16X67	127.09 (19.7)	W18X40	75.90 (11.76)
4	457X152X52 UB	66.5 (10.30)	W10X45	85.81 (13.3)	W21X44	83.70 (12.97)
5	457X152X60 UB	75.9 (11.76)	W14X68	129.03 (20.0)	W21X44	83.70 (12.97)
6	457X152X52 UB	66.5 (10.30)	W10X49	92.90 (14.4)	W16X45	86.00 (13.33)
7	457X152X52 UB	66.5 (10.30)	W12X45	85.16 (13.2)	W21X44	83.70 (12.97)
8	457X152X52 UB	66.5 (10.30)	W12X53	100.65 (15.6)	W21X44	83.70 (12.97)
9	203X203X60 UC	75.8 (11.75)	W14X132	250.32 (38.8)	W14X30	57.30 (8.88)
10	203X203X52 UC	66.4 (10.29)	W14X109	206.45 (32.0)	W14X30	57.30 (8.88)
11	254X254X73 UC	92.9 (14.40)	W12X96	181.94 (28.2)	W16X36	68.10 (10.56)
12	203X203X46 UC	58.8 (9.11)	W12X65	123.22 (19.1)	W16X31	58.90 (9.13)
13	254X254X73 UC	92.9 (14.40)	W10X49	92.90 (14.4)	W8X31	58.60 (9.08)
14	203X203X52 UC	66.4 (10.29)	W24X55	104.51 (16.2)	W14X34	64.50 (10.0)
15	305X305X97 UC	123.0 (19.07)	W21X44	83.87 (13.0)	W8X40	75.60 (11.72)
16	203X203X71 UC	91.1 (14.12)	W24X68	129.68 (20.1)	W16X40	75.80 (11.75)
17	305X305X118 UC	150.0 (23.25)	W24X55	104.51 (16.2)	W18X60	114.00 (17.67)
18	254X254X73 UC	92.9 (14.40)	W24X68	129.68 (20.1)	W14X43	81.40 (12.62)
19	305X305X118 UC	150.0 (23.25)	W24X62	117.42 (18.2)	W14X61	116.00 (17.98)
20	305X305X97 UC	123.0 (19.07)	W21X44	83.87 (13.0)	W14X48	91.10 (14.12)
21	356X368X153 UC	195.0 (30.22)	W24X68	129.68 (20.1)	W12X72	136.00 (21.08)
Weight	33011kg (72711.45 lb)		47472.66 kg (104659.29 lb)		32868.54 kg (72462.73 lb)	

BS5950 in spite of the fact that the drift limitations are H/400 in LRFD and H/300 in BS5950. The reason for this is that in the discrete set of steel sections of the optimum design due to BS5950 there are only 96 available British steel sections (64 Universal Beam sections and 32 Universal Column sections) while in the design due to LRFD there are 272 W-sections in the discrete list. Hence the optimum design that is based on LRFD specifications has larger design space to select from compare to the design space which makes use of British steel sections. This difference provides better selection possibilities to the algorithms based on LRFD.

6 Conclusions

Adaptive harmony search algorithm presented in this chapter is efficient and robust algorithm that can be employed with confidence in the optimum design of real size steel skeletal structures. In this technique the harmony search parameters are dynamically adjusted by the algorithm itself taking into account varying features of the design problem under consideration. The algorithm itself automatically changes the values of harmony considering rate (*hmcr*) and pitch adjustment rate (*par*) depending on the experience obtained through the design process. Hence, varying features of a design space are automatically accounted by the algorithm for establishing a tradeoff between explorative and exploitative search for the most successful optimization process. It is shown through the design examples considered in the optimum design of real size steel structures that the adaptive harmony search method demonstrates good performance compare to standard harmony search method. Inspection of the design history of 209-member industrial factory building clearly shows the better performance in the convergence rate of the adaptive harmony search method compare to standard harmony search method. The optimum designs of 568-member and 1860-member steel structures are obtained by the presented technique without any difficulty. Furthermore, comparison carried out among seven recently developed metaheuristic optimization techniques has shown that adaptive harmony search algorithm finds the second lightest frame among the minimum weights obtained by these seven metaheuristic algorithms considered in this study while the standard harmony search method attains the heaviest design. Finally, the adaptive harmony search algorithm eliminates the necessity of carrying out a sensitivity analysis with different values of harmony search parameters whenever a new design problem is to be undertaken. This makes the algorithm more general and applicable to the optimum design of large size real-world steel structures. It is also shown in the last design example that use of different design codes results in different optimum designs. The allowable stress design method naturally yields heavier design due to the fact that nowhere in the frame stresses are allowed to reach their yield values. The load and resistance factor design and British Standards 5950 which are based on the ultimate state design concept gives lighter optimum designs as expected.

Acknowledgements

Authors would like to express their thanks and appreciations to PhD students Ferhat Erdal, Serdar Carbas, Erkan Doğan and İbrahim Aydoğdu for their meticulous efforts in obtaining the optimum solutions of design examples considered in this chapter.

References

1. Horst, R., Pardolos, P.M. (eds.): Handbook of global optimization. Kluwer Academic Publishers, Dordrecht (1995)
2. Horst, R., Tuy, H.: Global optimization; Deterministic approaches. Springer, Heidelberg (1995)
3. Paton, R.: Computing with biological metaphors. Chapman and Hall, USA (1994)
4. Adami, C.: An introduction to artificial life. Springer, Heidelberg (1998)
5. Kochenberger, G.A., Glover, F.: Handbook of Metaheuristics. Kluwer Academic Publishers, Dordrecht (2003)
6. DeCastro, L.N., Von Zuben, F.J.: Recent developments in biologically inspired computing. Idea Group Publishing, USA (2005)
7. Dreo, J., Petrowski, A., Siarry, P., Taillard, E.: Metaheuristics for hard optimization. Springer, Heidelberg (2006)
8. Geem, Z.W., Kim, J.H.: A new heuristic optimization algorithm: harmony search. Simulation 76, 60–68 (2001)
9. Lee, K.S., Geem, Z.W.: A new structural optimization method based on harmony search algorithm. Computers and Structures 82, 781–798 (2004)
10. Lee, K.S., Geem, Z.W.: A new meta-heuristic algorithm for continuous engineering optimization: harmony search theory and practice. Computer Methods in Applied Mechanics and Engineering 194, 3902–3933 (2005)
11. AISC, Manual of steel construction-Allowable stress design (ASD). American Institutes of steel construction, Chicago, Illinois, USA (1989)
12. AISC, Manual of steel construction-Load and resistance factor design (LRFD). American Institutes of steel construction, Chicago, Illinois, USA (2000)
13. British Standards, BS5950, Structural use of steelworks in buildings, Part 1, Code of practice for design in simple and continuous construction, Hot Rolled Sections. British Standards Institution, London, UK (2000)
14. Dumonteil, P.: Simple equations for effective length factors. Engineering Journal, AISC 29(3), 111–115 (1992)
15. Hellesland, J.: Review and evaluation of effective length formulas. Research report, No. 94-2, University of Oslo, Sweden (1994)
16. ANSI/AISC 360-05, Specification for structural steel buildings. Chicago, Illinois, USA (2005)
17. McGuire, W.: Steel structures. Prentice-Hall, Englewood Cliffs (1968)
18. Değertekin, S.Ö.: Optimum design of steel frames using harmony search algorithm. Structural and Multidisciplinary Optimization 36, 393–401 (2008)

19. Saka, M.P.: Optimum design of steel swaying frames to BS5950 using harmony search algorithm. Journal of Constructional Steel Research, An International Journal 65(1), 36–43 (2009)
20. Erdal, F., Saka, M.P.: Effect of beam spacing in the harmony search based optimum design of grillages. Asian Journal of Civil Engineering 9(3), 215–228 (2008)
21. Saka, M.P., Erdal, F.: Harmony search based algorithm for the optimum design of grillage systems to LRFD-AISC. Structural and Multidisciplinary Optimization 38(1), 25–41 (2009)
22. Saka, M.P.: Optimum geometry design of geodesic domes using harmony search algorithm. Advances in Structural Engineering, An International Journal 10(6), 595–606 (2007)
23. Carbas, S., Saka, M.P.: Optimum design of single layer network domes using harmony search method. Asian Journal of Civil Engineering 10(1), 97–112 (2009)
24. Hasançebi, O., Çarbaş, S., Doğan, E., Erdal, F., Saka, M.P.: Performance evaluation of metaheuristic search techniques in the optimum design of real size pin jointed structures. Computers and Structures, An International Journal 87, 284–302 (2009)
25. Hasançebi, O., Çarbaş, S., Doğan, E., Erdal, F., Saka, M.P.: Optimum design of real size steel frames using non-deterministic search techniques. Computers and Structures, An International Journal (under review) (2009)
26. Hasançebi, O., Erdal, F., Saka, M.P.: An Adaptive Harmony Search Method for Structural Optimization. Journal of Structural Engineering, ASCE (under review) (2009)
27. Geem, Z.W., Kim, J.H., Loganathan, G.V.: Harmony search optimization: application to pipe network design. International Journal of Modeling and Simulation 22, 125–133 (2002)
28. Geem, Z.W.: Optimal cost design of water distribution networks using harmony search. Engineering Optimization 38, 259–280 (2006)
29. Geem, Z.W.: Improved harmony search from ensemble of music players. In: Gabrys, B., Howlett, R.J., Jain, L.C. (eds.) KES 2006. LNCS (LNAI), vol. 4251, pp. 86–93. Springer, Heidelberg (2006)
30. ASCE 7-05, Minimum design loads for building and other structures. American Society of Civil Engineering (2005)
31. Kameshki, E.S.: Comparison of BS 5950 and AISC-LRFD codes of practice. Practice Periodical on Structural design and Construction, ASCE 3(3), 105–118 (1998)

Harmony Particle Swarm Algorithm for Structural Design Optimization

Lijuan Li[1] and Feng Liu[2]

Abstract. This chapter introduces the application of an improved particle swarm algorithm to pin connected space structures. The algorithm is named harmony particle swarm optimization (HPSO), as it is based on harmony search schemes and the standard particle swarm algorithm. The efficiency of HPSO for pin connected structures with different variable types including continuous variables and discrete variables is compared with that of other intelligent algorithms, and the implementation of HPSO is presented in detail. An optimal result of a complex practical double-layer grid shell structure is presented to value the effectiveness of the HPSO.

1 Introduction

In the last 30 years, a great attention has been paid to structural optimization, since material consumption is one of the most important factors influencing building construction. Designers prefer to reduce the volume or weight of structures through optimization. Many traditional mathematical optimization algorithms have been used in structural optimization problems. The traditional optimal algorithms provide a useful strategy to obtain the global optimal solution in a simple model.

However, many practical engineering optimal problems are very complex and hard to solve by the traditional optimal algorithms. Recently, evolutionary algorithms (EAs), such as genetic algorithms (GAs), evolutionary programming (EP) and evolution strategies (ES) have become more attractive because they do not require conventional mathematical assumptions and thus possess better global search abilities than the conventional optimization algorithms [1]. For example, GAs have been applied for structural optimization problems [2-4].

A new evolutionary algorithm called particle swarm optimizer (PSO) was developed by Kennedy and Eberhart [5], which was inspired by the social

[1] Faculty of Civil and Transportation Engineering, Guangdong University of Technology, Guangzhou, China
lilj@gdut.edu.cn
[2] Faculty of Civil and Transportation Engineering, Guangdong University of Technology, Guangzhou, China
fliu@gdut.edu.cn

Z.W. Geem (Ed.): Harmony Search Algo. for Structural Design Optimization, SCI 239, pp. 121–157.
springerlink.com

behaviour of animals such as fish schooling and bird flocking. It is a population-based algorithm, which is based on the premise that social sharing of information among members of a species offers an evolutionary advantage. With respect to other algorithms such as evolutionary algorithms, a number of advantages make PSO an ideal candidate to be used in optimization tasks. The algorithm can handle continuous, discrete and integer variable types with ease. In addition, its easiness of implementation makes it more attractive for the applications of real-engineering optimization problems. Furthermore, it is a population-based algorithm, so it can be efficiently parallelized to reduce the total computational effort. The PSO has fewer parameters and is easier to implement than the GAs [6]. The PSO also shows a faster convergence rate than the other EAs for solving some optimization problems.

The foundation of PSO is based on the hypothesis that social sharing of information among conspecifics offers an evolutionary advantage. It involves a number of particles, which are initialized randomly in the search space of an objective function. These particles are referred to as swarm. Each particle of the swarm represents a potential solution of the optimization problem. The particles fly through the search space and their positions are updated based on the best positions of individual particles in each iteration. The objective function is evaluated for each particle and the fitness values of particles are obtained to determine which position in the search space is the best.

In each iteration, the swarm is updated using the following equations:

$$V_i^{k+1} = \omega V_i^k + c_1 r_1 \left(P_i^k - X_i^k \right) + c_2 r_2 \left(P_g^k - X_i^k \right) \tag{1}$$

$$X_i^{k+1} = X_i^k + V_i^{k+1} \tag{2}$$

where X_i and V_i represent the current position and the velocity of the ith particle respectively; P_i is the best previous position of the ith particle (called pbest) and P_g is the best global position among all the particles in the swarm (called gbest); r_1 and r_2 are two uniform random sequences generated from U(0, 1); and ω is the inertia weight used to discount the previous velocity of the particle persevered.

The PSO model is based on the following two factors:

(1) The autobiographical memory, which remembers the best previous position of each individual (P_i) in the swarm; and

(2) The publicized knowledge, which is the best solution (P_g) found currently by the population.

Geem [7] pointed out that although PSO may outperform other evolutionary algorithms in the early iterations, its performance may not be competitive as the problem size is increased. Recently, many investigations have been undertaken to improve the performance of the standard PSO (SPSO). He S et al. [8] found that adding the passive congregation model to the SPSO may increase its performance. Therefore, they improved the SPSO with passive congregation (PSOPC), which can improve the convergence rate and accuracy of the SPSO efficiently.

2 Constraint Handling Method: Fly-Back Mechanism

Most structural optimization problems include the problem-specific constraints, which are difficult to solve using the traditional mathematical optimization algorithms [9]. Penalty functions have been commonly used to deal with constraints. However, the major disadvantage of using the penalty functions is that some tuning parameters are added in the algorithm and the penalty coefficients have to be tuned in order to balance the objective and penalty functions. If appropriate penalty coefficients cannot be provided, difficulties will be encountered in the solution of the optimization problems [10, 11]. To avoid such difficulties, a new method, called 'fly-back mechanism', was developed.

For most of the optimization problems containing constraints, the global minimum locates on or close to the boundary of a feasible design space. The particles are initialized in the feasible region. When the optimization process starts, the particles fly in the feasible space to search the solution. If any one of the particles flies into the infeasible region, it will be forced to fly back to the previous position to guarantee a feasible solution. The particle which flies back to the previous position may be closer to the boundary at the next iteration. This makes the particles to fly to the global minimum in a great probability. Therefore, such a 'fly-back mechanism' technique is suitable for handling the optimization problem containing the constraints. Compared with the other constraint handling techniques, this method is relatively simple and easy to implement. Some experimental results have shown that it can find a better solution with a fewer iterations than the other techniques.

3 Harmony Particle Swarm Optimization (HPSO)

The harmony particle swarm optimizer (HPSO) [12] is based on the PSOPC and a harmony search (HS) scheme, and uses a 'fly-back mechanism' method to handle the constraints. The pseudo-code for the HPSO algorithm is listed in Table 1.

When a particle flies in the searching space, it may fly into infeasible regions. In this case, there are two possibilities. It may violate either the problem-specific constraints or the limits of the variables, as illustrated in Figure 1. Because the 'fly-back mechanism' technique is used to handle the problem-specific constraints, the particle will be forced to fly back to its previous position no matter if it violates the problem-specific constraints or the variable boundaries. If it flies out of the variable boundaries, the solution cannot be used even if the problem-specific constraints are satisfied. In our experiments, particles violate the variables' boundary frequently for some simple structural optimization problems. If the structure becomes complicated, the number of occurrences of violating tends to rise. In other words, a large amount of particles' flying behaviours are wasted, due to searching outside the variables' boundary. Although minimizing the maximum of the velocity can make fewer particles violate the variable boundaries, it may also prevent the particles to cross the problem-specific constraints. Therefore, we hope that all of the particles fly inside the variable boundaries and then to check whether they violate the problem-specific constraints and get better solutions or not. The particles, which fly outside the variables' boundary, have to be regenerated in an alternative way. Here, we introduce a new method to handle these particles. It is derived from one of the ideas in a new meta-heuristic algorithm called harmony search algorithm [13, 14].

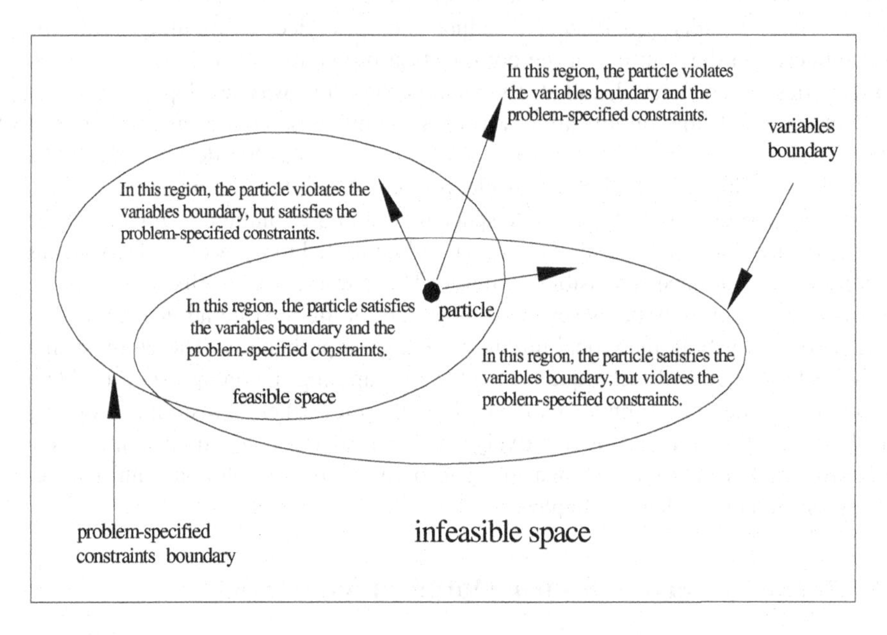

Fig. 1 The particle may violate the problem-specific constraints or the variables' boundary

Harmony search algorithm is based on natural musical performance processes that occur when a musician searches for a better state of harmony, such as during jazz improvisation [15]. The engineers seek for a global solution as determined by an objective function, just like the musicians seek to find musically pleasing harmony as determined by an aesthetic [16]. The harmony search algorithm includes a number of optimization operators, such as the harmony memory (HM), the harmony memory size (HMS), the harmony memory considering rate (HMCR), and the pitch adjusting rate (PAR). In this chapter, the harmony memory (HM) concept has been used in the PSO algorithm to avoid searching trapped in local solutions. The other operators have not been employed. How the HS algorithm generates a new vector from its harmony memory and how it is used to improve the PSO algorithm will be discussed as follows.

Table 1 The pseudo-code for the HPSO

```
Set k=1;
Randomly initialize positions and velocities of all particles;
    FOR (each particle i in the initial population)
        WHILE (the constraints are violated)
            Randomly re-generate the current particle X_i
        END WHILE
    END FOR
WHILE (the termination conditions are not met)
    FOR (each particle i in the swarm)
        Generate the velocity and update the position of the current particle (vector) X_i
        Check feasibility stage I: Check whether each component of the current vector
      violates its corresponding boundary or not. If it does, select the corresponding
      component of the vector from pbest swarm randomly.
        Check feasibility stage II: Check whether the current particle violates the problem
      specified constraints or not. If it does, reset it to the previous position X_ik-1.
        Calculate the fitness value f(X_ik) of the current particle.
        Update pbest: Compare the fitness value of pbest with f(X_ik). If the f(X_ik) is better than
      the fitness value of pbest, set pbest to the current position X_ik.
        Update gbest: Find the global best position in the swarm. If the f(X_ik) is better than the
      fitness value of gbest, gbest is set to the position of the current particle X_ik.
    END FOR
    Set k=k+1
END WHILE
```

In the HS algorithm, the harmony memory stores the feasible vectors, which are all in the feasible space. The harmony memory size determines how many vectors it stores. A new vector is generated by selecting the components of different vectors randomly in the harmony memory. Undoubtedly, the new vector does not violate the variables boundaries, but it is not certain if it violates the problem-specific constraints. When it is generated, the harmony memory will be updated by accepting this new vector if it gets a better solution and deleting the worst vector.

Similarly, the PSO stores the feasible and "good" vectors (particles) in the pbest swarm, as does the harmony memory in the HS algorithm. Hence, the vector (particle) violating the variables' boundaries can be generated randomly again by such a technique-selecting for the components of different vectors in the pbest swarm. There are two different ways to apply this technique to the PSO when any one of the components of the vector violates its corresponding variables' boundary. Firstly, all the components of this vector should be generated. Secondly, only this component of the vector should be generated again by such a technique. In our experiments, the results show that the former makes the particles moving to the local solution easily, and the latter can reach the global solution in relatively less number of iterations.

Therefore, applying such a technique to the PSOPC can improve its performance, although it already has a better convergence rate and accuracy than the PSO.

4 Application of the HPSO on Truss Structures with Continuous Variables

In this section, five pin-connected structures commonly used in literature are selected as benchmark problem to test the HPSO. The proposed algorithm is coded in FORTRAN language and executed on a Pentium 4, 2.93GHz machine.

The examples given in the simulation studies include

- a 10-bar planar truss structure subjected to four concentrated loads;
- a 17-bar planar truss structure subjected to a single concentrated load at its free end;
- a 22-bar spatial truss structure subjected to three load cases;
- a 25-bar spatial truss structure subjected to two load cases;
- a 72-bar spatial truss structure subjected to two load cases.

All these truss structures are analyzed by the finite element method (FEM).

The PSO, PSOPC and HPSO schemes are applied respectively to all these examples and the results are compared in order to evaluate the performance of the new algorithm. For all these algorithms, a population of 50 individuals is used; the inertia weight ω decrease linearly from 0.9 to 0.4; and the value of acceleration constants c_1 and c_2 are set to be the same and equal to 0.8. The passive congregation coefficient c_3 is given as 0.6 for the PSOPC [8] and the HPSO algorithms. The maximum number of iterations is limited to 3000. The maximum velocity is set as the difference between the upper bound and the lower bound of variables, which ensures that the particles are able to fly into the problem-specific constraints' region.

4.1 Numerical Examples

4.1.1 The 10-Bar Planar Truss Structure

The 10-bar truss structure, shown in Fig. 2, has previously been analyzed by many researchers, such as Lee [16], Schmit [17], Rizzi [18], and Li [19]. The material density is 0.1 lb/in^3 and the modulus of elasticity is 10,000 kilo-pounds per square inch (ksi). The members are subjected to the stress limits of ± 25 ksi. All nodes in both vertical and horizontal directions are subjected to the displacement limits of ± 2.0 in. There are 10 design variables in this example and the minimum permitted cross-sectional area of each member is 0.1 in^2. Two cases are considered: Case 1, P_1=100 kilo-pounds force (kips) and P_2=0; Case 2, P_1=150 kips and P_2=50 kips.

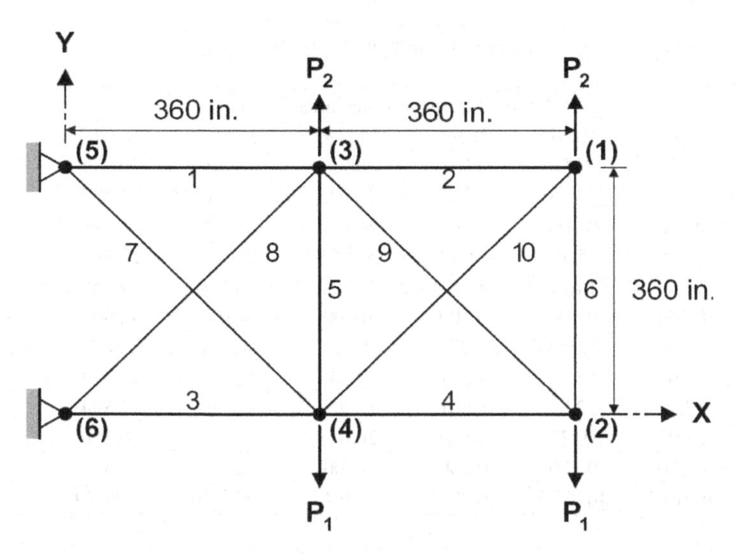

Fig. 2 A 10-bar planar truss structure

For both load cases, the PSOPC and the HPSO algorithms achieve the best solutions after 3,000 iterations. However, the latter is closer to the best solution than the former after about 500 iterations. The HPSO algorithm displays a faster convergence rate than the PSOPC algorithm in this example. The performance of the PSO algorithm is the worst among the three. Tables 2 and 3 show the solutions. Figs. 3 and 4 provide a comparison of the convergence rates of the three algorithms.

Table 2 Comparison of the designs for the 10-bar planar truss (Case 1)

Variables		Optimal cross-sectional areas (in.^{2})					
		Schmit [17]	Rizzi [18]	Lee [16]	Li [19] PSO	Li [19] PSOPC	Li [19] HPSO
1	A_1	33.43	30.73	30.15	33.469	30.569	30.704
2	A_2	0.100	0.100	0.102	0.110	0.100	0.100
3	A_3	24.26	23.93	22.71	23.177	22.974	23.167
4	A_4	14.26	14.73	15.27	15.475	15.148	15.183
5	A_5	0.100	0.100	0.102	3.649	0.100	0.100
6	A_6	0.100	0.100	0.544	0.116	0.547	0.551
7	A_7	8.388	8.542	7.541	8.328	7.493	7.460
8	A_8	20.74	20.95	21.56	23.340	21.159	20.978
9	A_9	19.69	21.84	21.45	23.014	21.556	21.508
10	A_{10}	0.100	0.100	0.100	0.190	0.100	0.100
Weight (lb)		5089.0	5076.66	5057.88	5529.50	5061.00	5060.92

Table 3 Comparison of the designs for the 10-bar planar truss (Case 2)

Variables		Optimal cross-sectional areas (in.^{2})					
		Schmit [17]	Rizzi [18]	Lee [16]	Li [19] PSO	Li [19] PSOPC	Li [19] HPSO
1	A_1	24.29	23.53	23.25	22.935	23.743	23.353
2	A_2	0.100	0.100	0.102	0.113	0.101	0.100
3	A_3	23.35	25.29	25.73	25.355	25.287	25.502
4	A_4	13.66	14.37	14.51	14.373	14.413	14.250
5	A_5	0.100	0.100	0.100	0.100	0.100	0.100
6	A_6	1.969	1.970	1.977	1.990	1.969	1.972
7	A_7	12.67	12.39	12.21	12.346	12.362	12.363
8	A_8	12.54	12.83	12.61	12.923	12.694	12.894
9	A_9	21.97	20.33	20.36	20.678	20.323	20.356
10	A_{10}	0.100	0.100	0.100	0.100	0.103	0.101
Weight (lb)		4691.84	4676.92	4668.81	4679.47	4677.70	4677.29

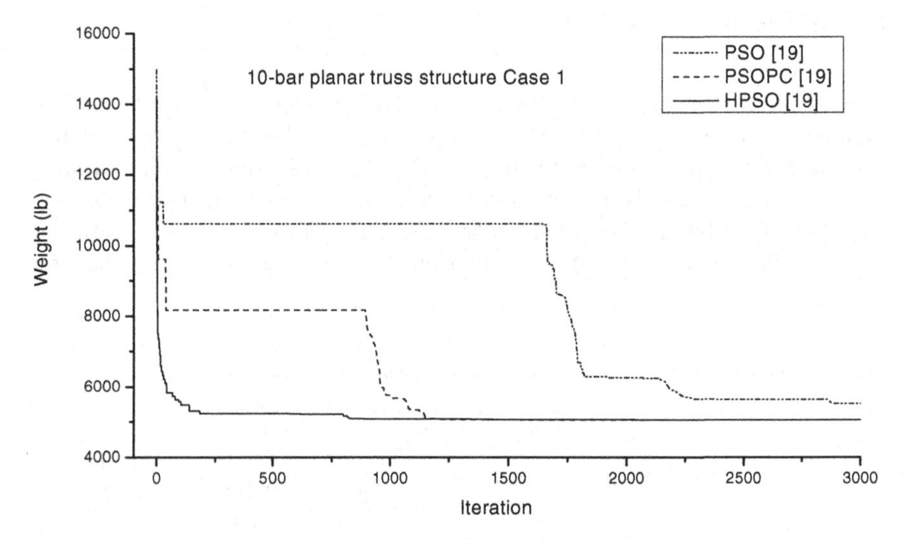

Fig. 3 Comparison of the convergence rates between the three algorithms for the 10-bar planar truss structure (Case 1)

4.1.2 The 17-Bar Planar Truss Structure

The 17-bar truss structure, shown in Fig. 5, had been analyzed by Khot [20], Adeli [21], Lee [16] and Li [19]. The material density is 0.268 lb/in.^{3} and the modulus of elasticity is 30,000 ksi. The members are subjected to the stress limits of ±50 ksi. All nodes in both directions are subjected to the displacement limits of ±2.0 in.

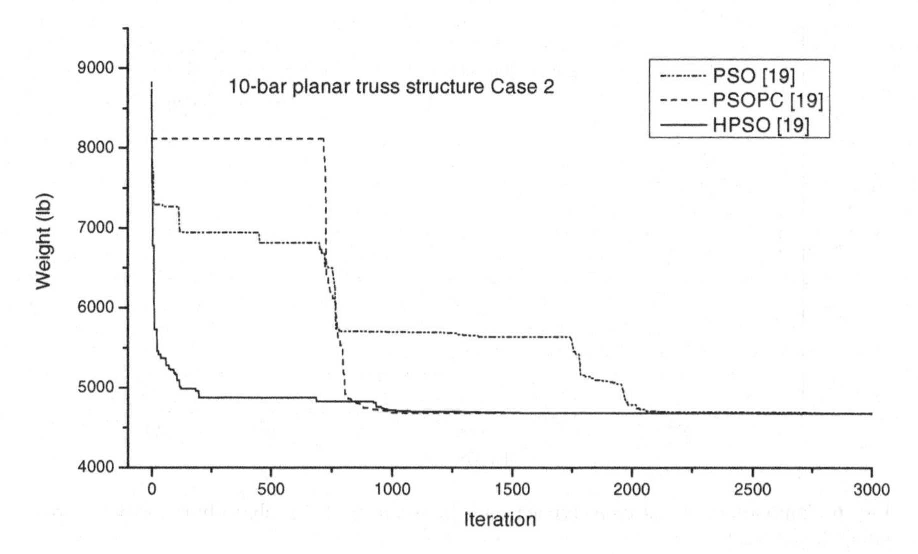

Fig. 4 Comparison of the convergence rates between the three algorithms for the 10-bar planar truss structure (Case 2)

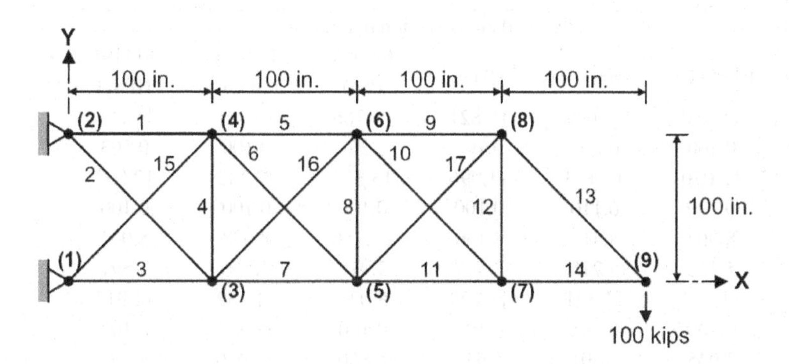

Fig. 5 A 17-bar planar truss structure

There are 17 design variables in this example and the minimum permitted cross-sectional area of each member is 0.1in.^{2}. A single vertical downward load of 100 kips at node 9 is considered. Table 4 shows the solutions and Fig. 6 compares the convergence rates of the three algorithms.

Both the PSOPC and HPSO algorithms achieve a good solution after 3,000 iterations and the latter shows a better convergence rate than the former, especially at the early stage of iterations. In this case, the PSO algorithm is not fully converged when the maximum number of iterations is reached.

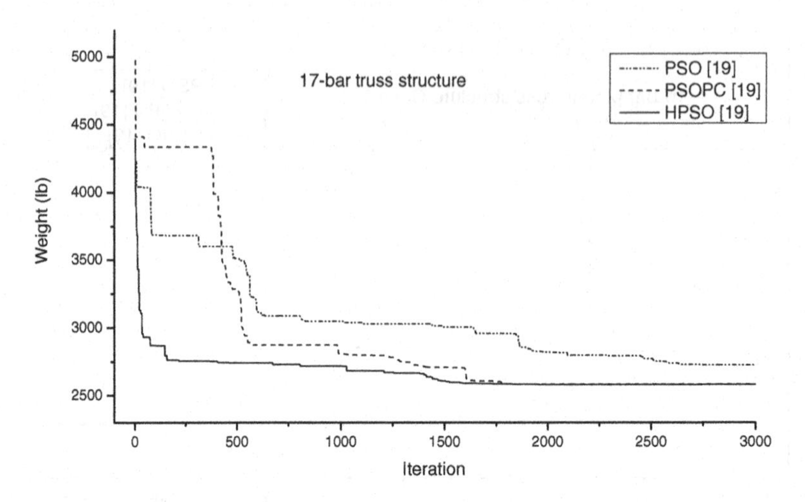

Fig. 6 Comparison of the convergence rates between the three algorithms for the 17-bar planar truss structure

Table 4 Comparison of the designs for the 17-bar planar truss

Variables		Optimal cross-sectional areas (in.^{2})					
		Khot [20]	Adeli [21]	Lee [16]	Li [19] PSO	Li [19] PSOPC	Li [19] HPSO
1	A_1	15.930	16.029	15.821	15.766	15.981	15.896
2	A_2	0.100	0.107	0.108	2.263	0.100	0.103
3	A_3	12.070	12.183	11.996	13.854	12.142	12.092
4	A_4	0.100	0.110	0.100	0.106	0.100	0.100
5	A_5	8.067	8.417	8.150	11.356	8.098	8.063
6	A_6	5.562	5.715	5.507	3.915	5.566	5.591
7	A_7	11.933	11.331	11.829	8.071	11.732	11.915
8	A_8	0.100	0.105	0.100	0.100	0.100	0.100
9	A_9	7.945	7.301	7.934	5.850	7.982	7.965
10	A_{10}	0.100	0.115	0.100	2.294	0.113	0.100
11	A_{11}	4.055	4.046	4.093	6.313	4.074	4.076
12	A_{12}	0.100	0.101	0.100	3.375	0.132	0.100
13	A_{13}	5.657	5.611	5.660	5.434	5.667	5.670
14	A_{14}	4.000	4.046	4.061	3.918	3.991	3.998
15	A_{15}	5.558	5.152	5.656	3.534	5.555	5.548
16	A_{16}	0.100	0.107	0.100	2.314	0.101	0.103
17	A_{17}	5.579	5.286	5.582	3.542	5.555	5.537
Weight (lb)		2581.89	2594.42	2580.81	2724.37	2582.85	2581.94

4.1.3 The 22-Bar Spatial Truss Structure

The 22-bar spatial truss structure, shown in Fig. 7, had been studied by Lee [16] and Li [19]. The material density is 0.1 lb/in.^{3} and the modulus of elasticity is 10,000 ksi. The stress limits of the members are listed in Table 5. All nodes in all three directions are subjected to the displacement limits of ±2.0 in. Three load cases are listed in Table 6. There are 22 members, which fall into 7 groups, as follows: (1) A_1~A_4, (2) A_5~A_6, (3) A_7~A_8, (4) A_9~A_{10}, (5) A_{11}~A_{14}, (6) A_{15}~A_{18}, and (7) A_{19}~A_{22}. The minimum permitted cross-sectional area of each member is 0.1 in.^{2}.

Table 5 Member stress limits for the 22-bar spatial truss structure

Variables		Compressive stress limitations (ksi)	Tensile stress Limitation (ksi)
1	A_1	24.0	36.0
2	A_2	30.0	36.0
3	A_3	28.0	36.0
4	A_4	26.0	36.0
5	A_5	22.0	36.0
6	A_6	20.0	36.0
7	A_7	18.0	36.0

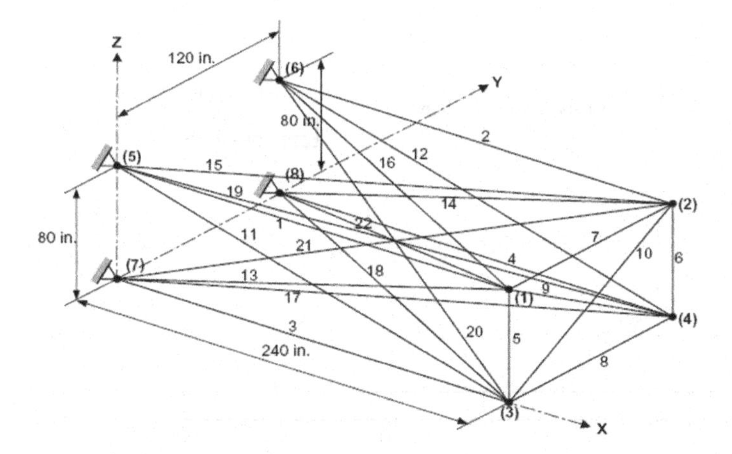

Fig. 7 A 22-bar spatial truss structure

In this example, the HPSO algorithm has converged after 50 iterations, while the PSOPC and PSO algorithms need more than 500 and 1000 iterations respectively. The optimum results obtained by using the HPSO algorithm are significantly better than that obtained by the HS and the PSO algorithms. Table 7 shows the optimal solutions of the four algorithms and Fig. 8 provides the convergence rates of three of the four algorithms.

Table 6 Load cases for the 22-bar spatial truss structure

Node	Case 1 (kips)			Case 2 (kips)			Case 3 (kips)		
	P_X	P_Y	P_Z	P_X	P_Y	P_Z	P_X	P_Y	P_Z
1	-20.0	0.0	-5.0	-20.0	-5.0	0.0	-20.0	0.0	35.0
2	-20.0	0.0	-5.0	-20.0	-50.0	0.0	-20.0	0.0	0.0
3	-20.0	0.0	-30.0	-20.0	-5.0	0.0	-20.0	0.0	0.0
4	-20.0	0.0	-30.0	-20.0	-50.0	0.0	-20.0	0.0	-35.0

Table 7 Comparison of the designs for the 22-bar spatial truss structure

Variables		Optimal cross-sectional areas (in.^{2})			
		Lee [16]	Li [19] PSO	Li [19] PSOPC	Li [19] HPSO
1	A_1	2.588	1.657	3.041	3.157
2	A_2	1.083	0.716	1.191	1.269
3	A_3	0.363	0.919	0.985	0.980
4	A_4	0.422	0.175	0.105	0.100
5	A_5	2.827	4.576	3.430	3.280
6	A_6	2.055	3.224	1.543	1.402
7	A_7	2.044	0.450	1.138	1.301
Weight (lb)		1022.23	1057.14	977.80	977.81

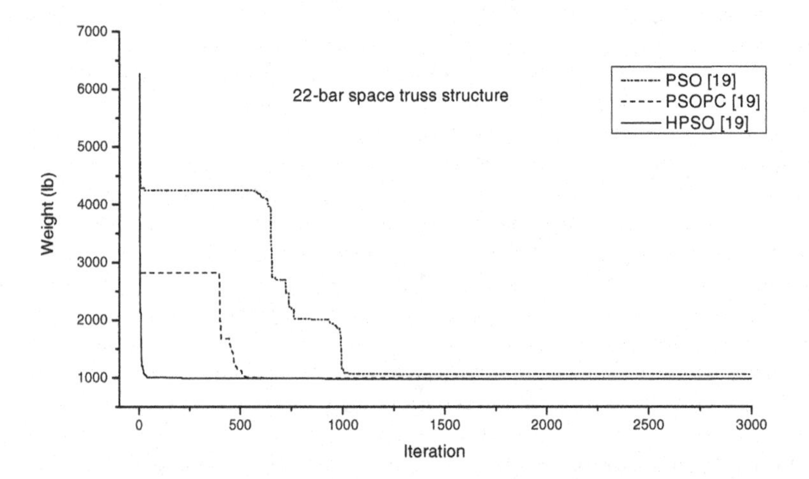

Fig. 8 Comparison of the convergence rates between the three algorithms for the 22-bar spatial truss structure

4.1.4 The 25-Bar Spatial Truss Structure

The 25-bar spatial truss structure shown in Fig. 9 had been studied by several researchers, such as Schmit [17], Rizzi [18], Lee [16] and Li [19]. The material density is 0.1 lb/in.^{3} and the modulus of elasticity is 10,000 ksi. The stress limits of the members are listed in Table 8. All nodes in all directions are subjected to the displacement limits of ±0.35 in. Two load cases listed in Table 9 are considered. There are 25 members, which are divided into 8 groups, as follows: (1) A_1, (2) A_2~A_5, (3) A_6~A_9, (4) A_{10}~A_{11}, (5) A_{12}~A_{13}, (6) A_{14}~A_{17}, (7) A_{18}~A_{21} and (8) A_{22}~A_{25}. The minimum permitted cross-sectional area of each member is 0.01 in^2.

For this spatial truss structure, it takes about 1000 and 3000 iterations, respectively, for the PSOPC and the PSO algorithms to converge. However the HPSO algorithm takes only 50 iterations to converge. Indeed, in this example, the PSO algorithm did not fully converge when the maximum number of iterations is reached. Table 10 shows the solutions and Fig. 10 compares the convergence rate of the three algorithms.

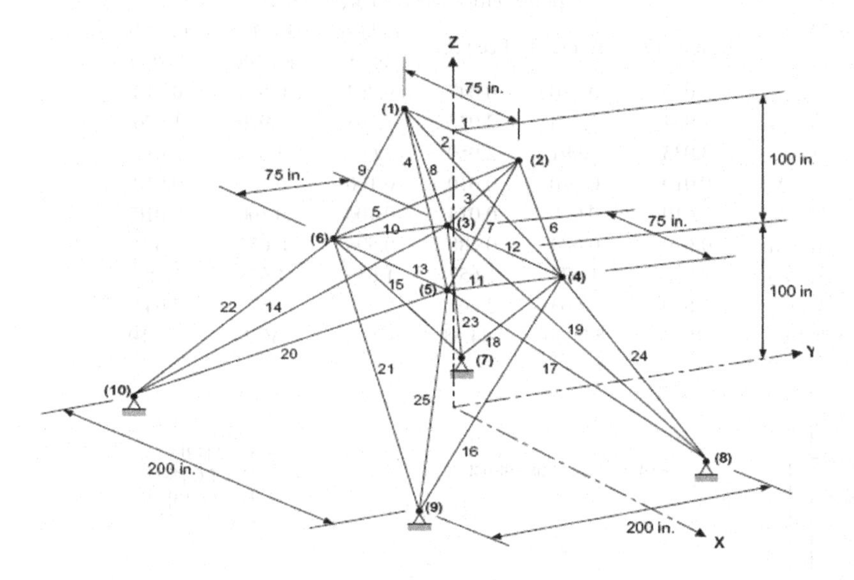

Fig. 9 A 25-bar spatial truss structure

Table 8 Member stress limits for the 25-bar spatial truss structure

Variables		Compressive stress limitations (ksi)	Tensile stress limitation (ksi)
1	A_1	35.092	40.0
2	A_2	11.590	40.0
3	A_3	17.305	40.0
4	A_4	35.092	40.0
5	A_5	35.902	40.0
6	A_6	6.759	40.0
7	A_7	6.959	40.0
8	A_8	11.802	40.0

Table 9 Load cases for the 25-bar spatial truss structure

Node	Case 1			Case 2		
	P_X (kips)	P_Y (kips)	P_Z (kips)	P_X (kips)	P_Y (kips)	P_Z (kips)
1	0.0	20.0	-5.0	1.0	10.0	-5.0
2	0.0	-20.0	-5.0	0.0	10.0	-5.0
3	0.0	0.0	0.0	0.5	0.0	0.0
6	0.0	0.0	0.0	0.5	0.0	0.0

Table 10 Comparison of the designs for the 25-bar spatial truss structure

Variables		Optimal cross-sectional areas (in.^{2})					
		Schmit [17]	Rizzi [18]	Lee [16]	Li [19] PSO	Li [19] PSOPC	Li [19] HPSO
1	A_1	0.010	0.010	0.047	9.863	0.010	0.010
2	A_2~A_5	1.964	1.988	2.022	1.798	1.979	1.970
3	A_6~A_9	3.033	2.991	2.950	3.654	3.011	3.016
4	A_{10}~A_{11}	0.010	0.010	0.010	0.100	0.100	0.010
5	A_{12}~A_{13}	0.010	0.010	0.014	0.100	0.100	0.010
6	A_{14}~A_{17}	0.670	0.684	0.688	0.596	0.657	0.694
7	A_{18}~A_{21}	1.680	1.677	1.657	1.659	1.678	1.681
8	A_{22}~A_{25}	2.670	2.663	2.663	2.612	2.693	2.643
Weight (lb)		545.22	545.36	544.38	627.08	545.27	545.19

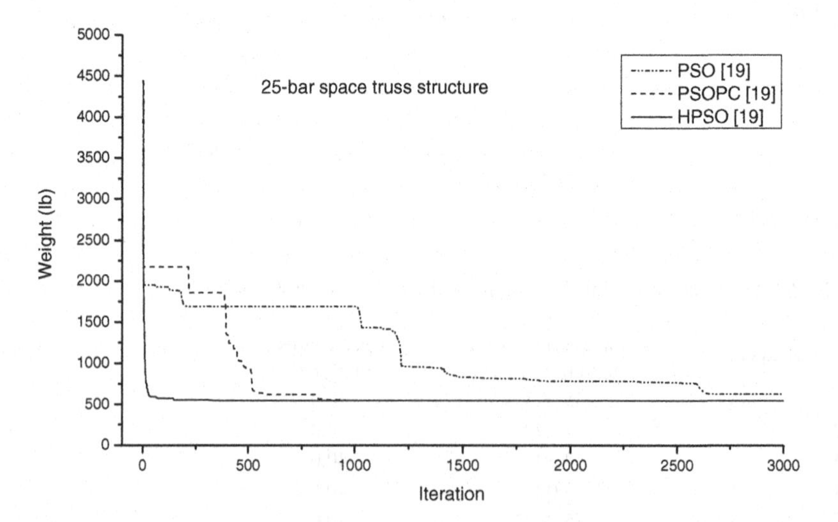

Fig. 10 Convergence rate comparison between the three algorithms for the 25-bar spatial truss structure

4.1.5 The 72-Bar Spatial Truss Structure

The 72-bar spatial truss structure shown in Fig. 11 had also been studied by many researchers, such as Schmit [17], Khot [20], Adeli [21], Lee [16], Sarma [22] and Li [19]. The material density is 0.1 lb/in.^{3} and the modulus of elasticity is 10,000 ksi. The members are subjected to the stress limits of ±25 ksi. The uppermost nodes are subjected to the displacement limits of ±0.25 in. in both the x and y directions. Two load cases are listed in Table 11. There are 72 members classified into 16 groups: (1) A_1~A_4, (2) A_5~A_{12}, (3) A_{13}~A_{16}, (4) A_{17}~A_{18}, (5) A_{19}~A_{22}, (6) A_{23}~A_{30} (7) A_{31}~A_{34}, (8) A_{35}~A_{36}, (9) A_{37}~A_{40}, (10) A_{41}~A_{48}, (11) A_{49}~A_{52}, (12) A_{53}~A_{54}, (13) A_{55}~A_{58}, (14) A_{59}~A_{66} (15) A_{67}~A_{70}, (16) A_{71}~A_{72}. For case 1, the minimum permitted cross-sectional area of each member is 0.1 in^2. For case 2, the minimum permitted cross-sectional area of each member is 0.01 in^2.

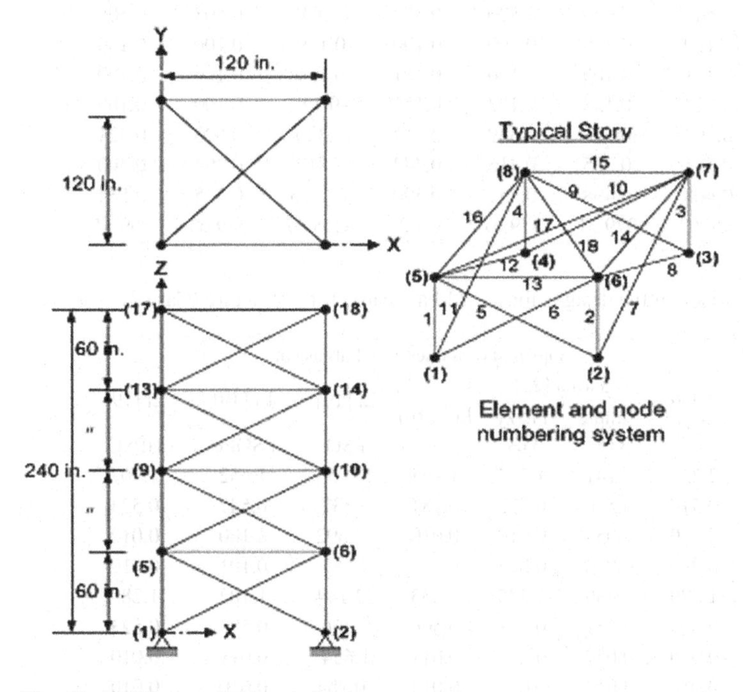

Fig. 11 A 72-bar spatial truss structure

Table 11 Load cases for the 72-bar spatial truss structure

Node	Case 1			Case 2		
	P_X (kips)	P_Y (kips)	P_Z (kips)	P_X (kips)	P_Y (kips)	P_Z (kips)
17	5.0	5.0	-5.0	0.0	0.0	-5.0
18	0.0	0.0	0.0	0.0	0.0	-5.0
19	0.0	0.0	0.0	0.0	0.0	-5.0
20	0.0	0.0	0.0	0.0	0.0	-5.0

Table 12 Comparison of the designs for the 72-bar spatial truss structure (Case 1)

Variables		Optimal cross-sectional areas (in.^{2})						
		Schmit [17]	Adeli [21]	Khot [20]	Lee [16]	Li [19] PSO	Li [19] PSOPC	Li [19] HPSO
1	$A_1 \sim A_4$	2.078	2.026	1.893	1.7901	41.794	1.855	1.857
2	$A_5 \sim A_{12}$	0.503	0.533	0.517	0.521	0.195	0.504	0.505
3	$A_{13} \sim A_{16}$	0.100	0.100	0.100	0.100	10.797	0.100	0.100
4	$A_{17} \sim A_{18}$	0.100	0.100	0.100	0.100	6.861	0.100	0.100
5	$A_{19} \sim A_{22}$	1.107	1.157	1.279	1.229	0.438	1.253	1.255
6	$A_{23} \sim A_{30}$	0.579	0.569	0.515	0.522	0.286	0.505	0.503
7	$A_{31} \sim A_{34}$	0.100	0.100	0.100	0.100	18.309	0.100	0.100
8	$A_{35} \sim A_{36}$	0.100	0.100	0.100	0.100	1.220	0.100	0.100
9	$A_{37} \sim A_{40}$	0.264	0.514	0.508	0.517	5.933	0.497	0.496
10	$A_{41} \sim A_{48}$	0.548	0.479	0.520	0.504	19.545	0.508	0.506
11	$A_{49} \sim A_{52}$	0.100	0.100	0.100	0.100	0.159	0.100	0.100
12	$A_{53} \sim A_{54}$	0.151	0.100	0.100	0.101	0.151	0.100	0.100
13	$A_{55} \sim A_{58}$	0.158	0.158	0.157	0.156	10.127	0.100	0.100
14	$A_{59} \sim A_{66}$	0.594	0.550	0.539	0.547	7.320	0.525	0.524
15	$A_{67} \sim A_{70}$	0.341	0.345	0.416	0.442	3.812	0.394	0.400
16	$A_{71} \sim A_{72}$	0.608	0.498	0.551	0.590	18.196	0.535	0.534
Weight (lb)		388.63	379.31	379.67	379.27	6818.67	369.65	369.65

Table 13 Comparison of the designs for the 72-bar spatial truss structure (Case 2)

Variables		Optimal cross-sectional areas (in.^{2})						
		Adeli [21]	Sarma [22] Simple GA	Sarma [22] Fuzzy GA	Lee [16]	Li [19] PSO	Li [19] PSOPC	Li [19] HPSO
1	$A_1 \sim A_4$	2.755	2.141	1.732	1.963	40.053	1.652	1.907
2	$A_5 \sim A_{12}$	0.510	0.510	0.522	0.481	0.237	0.547	0.524
3	$A_{13} \sim A_{16}$	0.010	0.054	0.010	0.010	21.692	0.100	0.010
4	$A_{17} \sim A_{18}$	0.010	0.010	0.013	0.011	0.657	0.101	0.010
5	$A_{19} \sim A_{22}$	1.370	1.489	1.345	1.233	22.144	1.102	1.288
6	$A_{23} \sim A_{30}$	0.507	0.551	0.551	0.506	0.266	0.589	0.523
7	$A_{31} \sim A_{34}$	0.010	0.057	0.010	0.011	1.654	0.011	0.010
8	$A_{35} \sim A_{36}$	0.010	0.013	0.013	0.012	10.284	0.010	0.010
9	$A_{37} \sim A_{40}$	0.481	0.565	0.492	0.538	0.559	0.581	0.544
10	$A_{41} \sim A_{48}$	0.508	0.527	0.545	0.533	12.883	0.458	0.528
11	$A_{49} \sim A_{52}$	0.010	0.010	0.066	0.010	0.138	0.010	0.019
12	$A_{53} \sim A_{54}$	0.643	0.066	0.013	0.167	0.188	0.152	0.020
13	$A_{55} \sim A_{58}$	0.215	0.174	0.178	0.161	29.048	0.161	0.176
14	$A_{59} \sim A_{66}$	0.518	0.425	0.524	0.542	0.632	0.555	0.535
15	$A_{67} \sim A_{70}$	0.419	0.437	0.396	0.478	3.045	0.514	0.426
16	$A_{71} \sim A_{72}$	0.504	0.641	0.595	0.551	1.711	0.648	0.612
Weight (lb)		376.50	372.40	364.40	364.33	5417.02	368.45	364.86

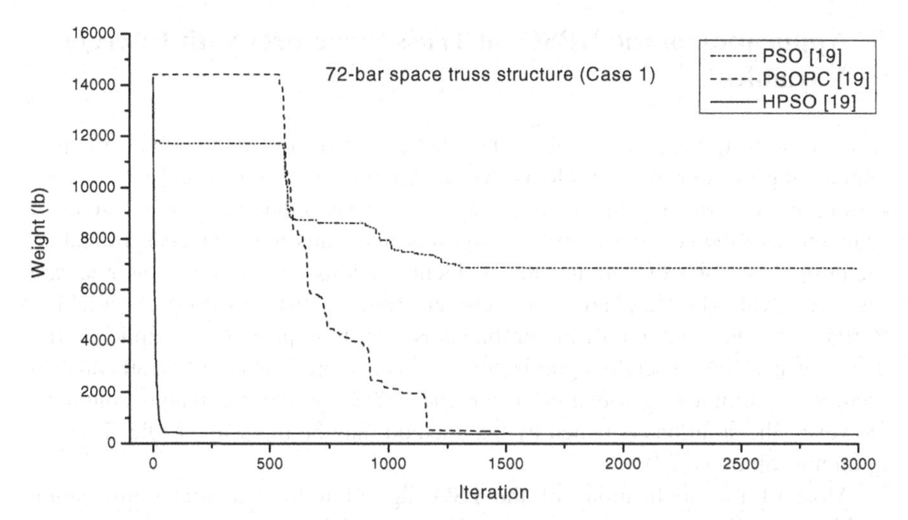

Fig. 12 Comparison of the convergence rates between the three algorithms for the 72-bar spatial truss structure (Case 1)

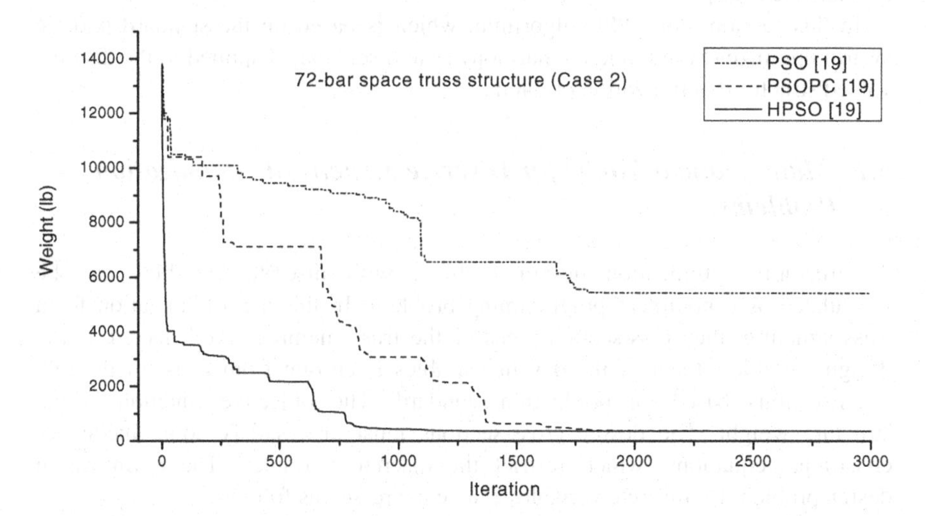

Fig. 13 Comparison of the convergence rates between the three algorithms for the 72-bar spatial truss structure (Case 2)

For both the loading cases, the PSOPC and the HPSO algorithms can achieve the optimal solution after 2500 iterations. However, the latter shows a faster convergence rate than the former, especially at the early stage of iterations. The PSO algorithm cannot reach the optimal solution after the maximum number of iterations. The solutions of the two loading cases are given in Tables 12 and 13 respectively. Figs. 12 and 13 compare the convergence rate of the three algorithms for the two loading cases.

5 Application of the HPSO on Truss Structures with Discrete Variables

In the past thirty years, many algorithms have been developed to solve structural engineering optimization problems. Most of these algorithms are based on the assumption that the design variables are continuously valued and the gradients of functions and the convexity of the design problem satisfied. However, in reality, the design variables of optimization problems such as the cross-section areas are discretely valued. They are often chosen from a list of discrete variables. Furthermore, the function of the problems is hard to express in an explicit form. Traditionally, the discrete optimization problems are solved by mathematical methods by employing round-off techniques based on the continuous solutions. However, the solutions obtained by this method may be infeasible or far from the optimum solutions [23].

Most of the applications of the PSO algorithm to structural optimization problems are based on the assumption that the variables are continuous. Only in few papers PSO algorithm is used to solve the discrete structural optimization problems [24, 25].

In this section, the HPSO algorithm, which is based on the standard particle swarm optimize (SPSO) and the harmony search scheme, is applied to the discrete valued structural optimization problems.

5.1 Mathematical Model for Discrete Structural Optimization Problems

A structural optimization design problem with discrete variables can be formulated as a nonlinear programming problem. In the size optimization for a truss structure, the cross-section areas of the truss members are selected as the design variables. Each of the design variables is chosen from a list of discrete cross-sections based on production standard. The objective function is the structure weight. The design cross-sections must also satisfy some inequality constraints equations, which restrict the discrete variables. The optimization design problem for discrete variables can be expressed as follows:

$$\min f\left(x^1, x^2, ..., x^d\right)$$

subject to

$$\begin{gathered} g_q\left(x^1, x^2, ..., x^d\right) \le 0 \\ x^d \in S_d = \left\{X_1, X_2, \cdots, X_p\right\} \\ d = 1, 2, \cdots, D \\ q = 1, 2, \cdots, M \end{gathered} \tag{3}$$

where $f\left(x^1, x^2, ..., x^d\right)$ is the truss's weight function, which is a scalar function. And $x^1, x^2, ..., x^d$ represent a set of design variables. The design variable x^d belongs to a scalar S_d, which includes all permissive discrete variables $\{X_1, X_2, ... X_p\}$. The inequality $g_q\left(x^1, x^2, ..., x^d\right) \leq 0$ represents the constraint functions. The letters D and M are the number of the design variables and inequality functions respectively. The letter p is the number of available variables.

5.2 The Harmony Particle Swarm Optimizer (HPSO) for Discrete Variables

The harmony particle swarm optimizer (HPSO) algorithm introduced by Li [19] is originally applied to continuous variable optimization problems. The HPSO algorithm then was used for discrete problems [26]. Similarly, The HPSO algorithm for the discrete valued variables can be expressed as follows:

$$V_i^{(k+1)} = \omega V_i^{(k)} + c_1 r_1 \left(P_i^{(k)} - x_i^{(k)}\right) + c_2 r_2 \left(P_g^{(k)} - x_i^{(k)}\right) + c_3 r_3 \left(R_i^{(k)} - x_i^{(k)}\right) \tag{4}$$

$$x_i^{(k+1)} = INT\left(x_i^{(k)} + V_i^{(k+1)}\right) \quad 1 \leq i \leq n \tag{5}$$

where $\boldsymbol{x}_i$ is the vector of a particle's position, and x_i^d is one component of this vector. After the (k+1)th iterations, if $x_i^d < x^d(LowerBound)$ or $x_i^d > x^d(UpperBound)$, the scalar x_i^d is regenerated by selecting the corresponding component of the vector from pbest swarm randomly, which can be described as follows:

$$x_i^d = \left(P_b\right)_t^d, \; t = INT\left(rand\left(1, n\right)\right) \tag{6}$$

where $\left(P_b\right)_t^d$ denotes the dth dimension scalar of pbest swarm of the tth particle, and t denotes a random integer number.

In this section, the HPSO algorithm is tested by five truss structures. The algorithm proposed is coded in FORTRAN language and executed on a Pentium 4, 2.93GHz machine.

The PSO, the PSOPC and the HPSO algorithms for discrete variables are applied to all these examples and the results are compared in order to evaluate the performance of the HPSO algorithm for discrete variables. For all these algorithms, a population of 50 individuals are used, the inertia weight ω, which starts at 0.9 and ends at 0.4, decreases linearly, and the value of acceleration constants c_1 and c_2 are set to 0.5 [27]. The passive congregation coefficient c_3 is set to 0.6 for the PSOPC and the HPSO algorithms. All these truss structures have been analyzed by the finite element method (FEM). The maximum velocity is set as the difference between the upper and the lower bounds, which ensures that the particles are able to fly across the problem-specific constraints' region. Different iteration numbers are used for different optimization structures, with smaller iteration number for smaller variable number structures and larger one for large variable number structures.

5.3 Numerical Examples

5.3.1 A 10-Bar Planar Truss Structure

A 10-bar truss structure, shown in Fig. 14, has previously been analyzed by many researchers, such as Wu [24], Rajeev [28], Ringertz [29] and Li [26]. The material density is 0.1lb/in^3 and the modulus of elasticity is 10,000 ksi. The members are subjected to stress limitations of $\pm$25 ksi. All nodes in both directions are subjected to displacement limitations of $\pm$2.0 in. P_1=105 lbs, P_2=0. There are 10 design variables and two load cases in this example to be optimized. For case 1: the discrete variables are selected from the set D={1.62, 1.80, 1.99, 2.13, 2.38, 2.62, 2.63, 2.88, 2.93, 3.09, 3.13, 3.38, 3.47, 3.55, 3.63, 3.84, 3.87, 3.88, 4.18, 4.22, 4.49, 4.59, 4.80, 4.97, 5.12, 5.74, 7.22, 7.97, 11.50, 13.50, 13.90, 14.20, 15.50, 16.00, 16.90, 18.80, 19.90, 22.00, 22.90, 26.50, 30.00, 33.50} (in^2) ; For case 2: the discrete variables are selected from the set D={0.1, 0.5, 1.0, 1.5, 2.0, 2.5, 3.0, 3.5, 4.0, 4.5, 5.0, 5.5, 6.0, 6.5, 7.0, 7.5, 8.0, 8.5, 9.0, 9.5, 10.0, 10.5, 11.0,

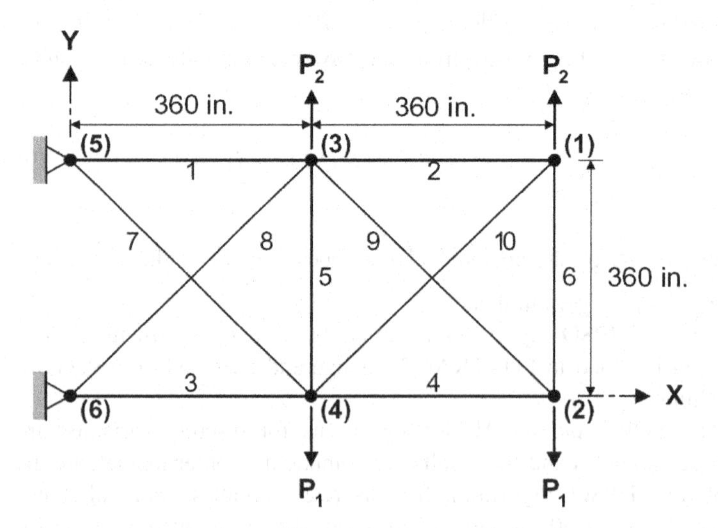

Fig. 14 A 10-bar planar truss structure

11.5, 12.0, 12.5, 13.0, 13.5, 14.0, 14.5, 15.0, 15.5, 16.0, 16.5, 17.0, 17.5, 18.0, 18.5, 19.0, 19.5, 20.0, 20.5, 21.0, 21.5, 22.0, 22.5, 23.0, 23.5, 24.0, 24.5, 25.0, 25.5, 26.0, 26.5, 27.0, 27.5, 28.0, 28.5, 29.0, 29.5, 30.0, 30.5, 31.0, 31.5} (in^2). A maximum number of 1000 iterations is imposed.

Table 14 and Table 15 give the comparison of optimal design results for the 10-bar planar truss structure under two load cases respectively. Fig.15 and Fig.16 show the comparison of convergence rates for the 10-bar truss structure. From the Table 14 and Table 15, we find the results obtained by the HPSO algorithm are larger than those of Wu's [24]. However, it is found that Wu's results do not satisfy the constraints of this problem. It is believed that Wu's results need to be further evaluated. For both cases of this structure, the PSO, PSOPC and HPSO algorithms have achieved the optimal solutions after 1,000 iterations. But the latter is much closer to the best solution than the former in the early iterations.

Table 14 Comparison of optimal designs for the 10-bar planar truss structure (case 1)

Variables (in^2)	Wu [24]	Rajeev [28]	Li [26] PSO	Li [26] PSOPC	Li [26] HPSO
A_1	26.50	33.50	30.00	30.00	30.00
A_2	1.62	1.62	1.62	1.80	1.62
A_3	16.00	22.00	30.00	26.50	22.90
A_4	14.20	15.50	13.50	15.50	13.50
A_5	1.80	1.62	1.62	1.62	1.62
A_6	1.62	1.62	1.80	1.62	1.62
A_7	5.12	14.20	11.50	11.50	7.97
A_8	16.00	19.90	18.80	18.80	26.50
A_9	18.80	19.90	22.00	22.00	22.00
A_{10}	2.38	2.62	1.80	3.09	1.80
Weight (lb)	4376.20	5613.84	5581.76	5593.44	5531.98

Table 15 Comparison of optimal designs for the 10-bar planar truss structure (case 2)

Variables (in^2)	Wu [24]	Ringertz [29]	Li [26] PSO	Li [26] PSOPC	Li [26] HPSO
A1	30.50	30.50	24.50	25.50	31.50
A2	0.50	0.10	0.10	0.10	0.10
A3	16.50	23.00	22.50	23.50	24.50
A4	15.00	15.50	15.50	18.50	15.50
A5	0.10	0.10	0.10	0.10	0.10
A6	0.10	0.50	1.50	0.50	0.50
A7	0.50	7.50	8.50	7.50	7.50
A8	18.00	21.0	21.50	21.50	20.50
A9	19.50	21.5	27.50	23.50	20.50
A10	0.50	0.10	0.10	0.10	0.10
Weight (lb)	4217.30	5059.9	5243.71	5133.16	5073.51

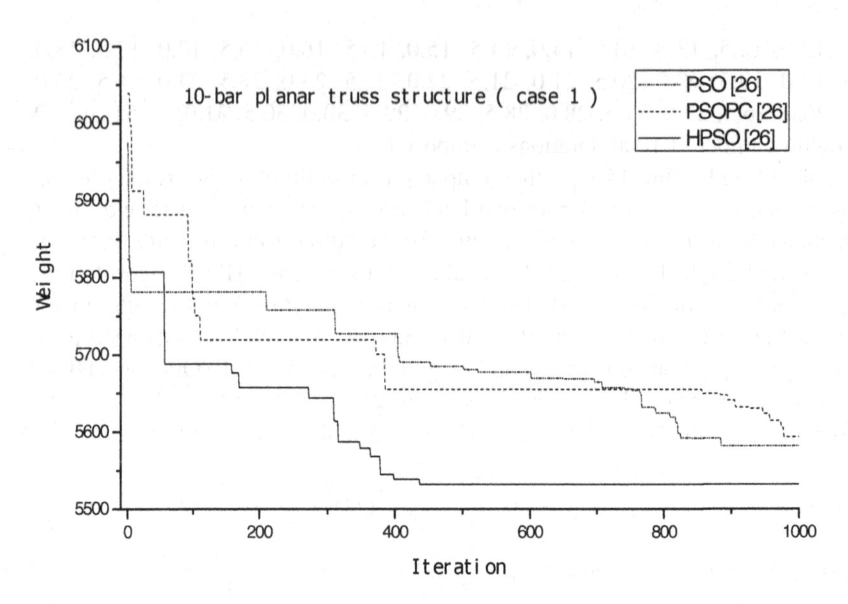

Fig. 15 Comparison of convergence rates for the 10-bar planar truss structure (Case 1)

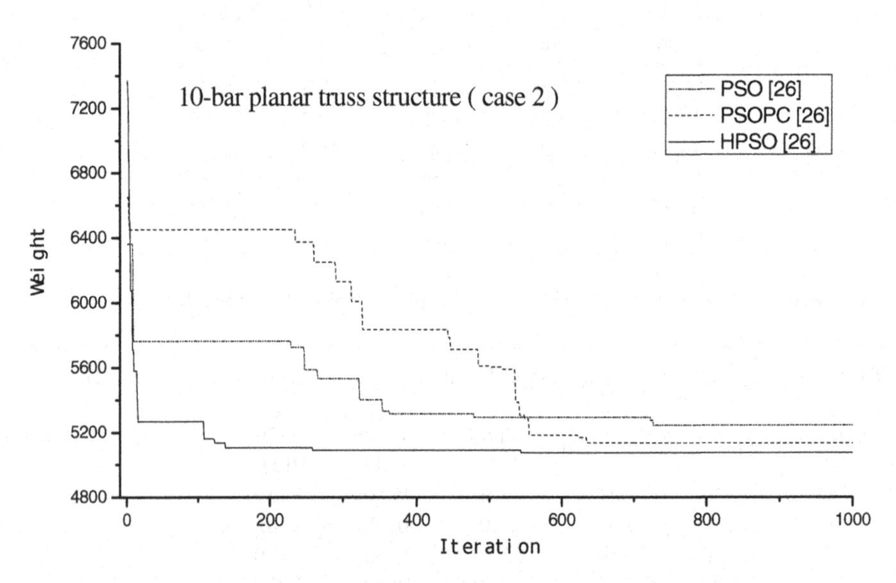

Fig. 16 Comparison of convergence rates for the 10-bar planar truss structure (Case 2)

5.3.2 A 15-Bar Planar Truss Structure

A 15-bar planar truss structure, shown in Fig. 17, has previously been analyzed by Zhang [30] and Li [26]. The material density is 7800kg/m^3 and the modulus of elasticity is 200GPa. The members are subjected to stress limitations of ±120MPa. All nodes in both directions are subjected to displacement limitations of ±10mm. There are 15 design variables in this example. The discrete variables are selected from the set D= {113.2, 143.2, 145.9, 174.9, 185.9, 235.9, 265.9, 297.1, 308.6, 334.3, 338.2, 497.8, 507.6, 736.7, 791.2, 1063.7} (mm^2). Three load cases are considered: Case 1: P_1=35kN, P_2=35kN, P_3=35kN; Case 2: P_1=35kN, P_2=0kN, P_3=35kN; Case 3: P_1=35kN, P_2=35kN, P_3=0kN. A maximum number of 500 iterations is imposed.

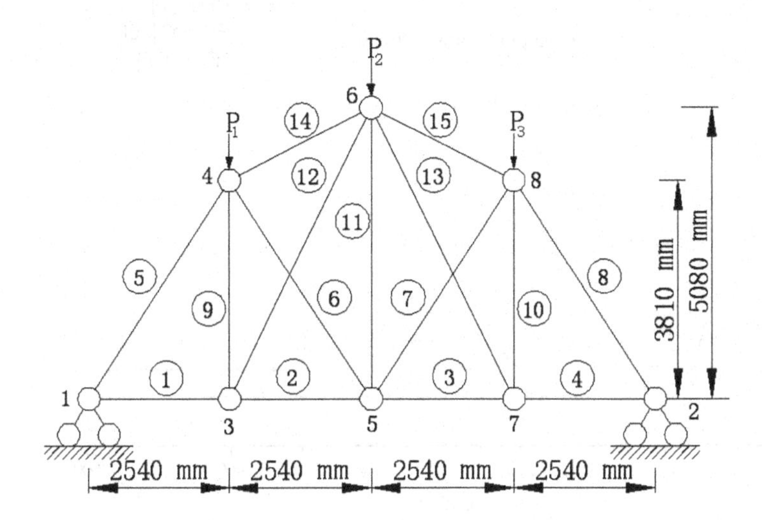

Fig. 17 A 15-bar planar truss structure

Table 16 Comparison of optimal designs for the 15-bar planar truss structure

Variables (mm^2)	Zhang [30]	Li [26] PSO	Li [26] PSOPC	Li [26] HPSO
A_1	308.6	185.9	113.2	113.2
A_2	174.9	113.2	113.2	113.2
A_3	338.2	143.2	113.2	113.2
A_4	143.2	113.2	113.2	113.2
A_5	736.7	736.7	736.7	736.7
A_6	185.9	143.2	113.2	113.2
A_7	265.9	113.2	113.2	113.2
A_8	507.6	736.7	736.7	736.7
A_9	143.2	113.2	113.2	113.2
A_{10}	507.6	113.2	113.2	113.2
A_{11}	279.1	113.2	113.2	113.2
A_{12}	174.9	113.2	113.2	113.2
A_{13}	297.1	113.2	185.9	113.2
A_{14}	235.9	334.3	334.3	334.3
A_{15}	265.9	334.3	334.3	334.3
Weight (kg)	142.117	108.84	108.96	105.735

Table 16 and Fig. 18 give the comparison of optimal design results and convergence rates of 15-bar planar truss structure respectively. It can be seen that, after 500 iterations, three algorithms have obtained good results, which are better than the Zhang's. The Fig. 18 shows that the HPSO algorithm has the fastest convergence rate, especially in the early iterations.

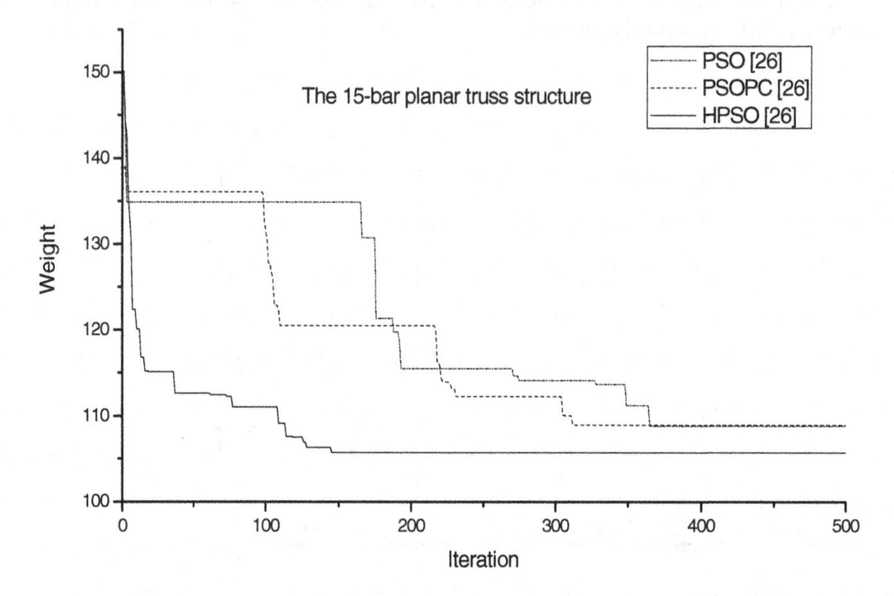

Fig. 18 Comparison of convergence rates for the 15-bar planar truss structure

5.3.3 A 25-Bar Spatial Truss Structure

A 25-bar spatial truss structure, shown in Fig. 19, has been studied by Wu [24], Rajeev [28], Ringertz [29], Lee [13] and Li [26]. The material density is 0.1 lb/in.^{3} and the modulus of elasticity is 10,000 ksi. The stress limitations of the members are $\pm$40000psi. All nodes in three directions are subjected to displacement limitations of $\pm$0.35 in. The structure includes 25 members, which are divided into 8 groups, as follows: (1) A_1, (2) A_2~A_5 (3) A_6~A_9, (4) A_{10}~A_{11}, (5) A_{12}~A_{13}, (6) A_{14}~A_{17}, (7) A_{18}~A_{21} and (8) A_{22}~A_{25}. There are three optimization cases to be

Table 17 The load case 1 for the 25-bar spatial truss structure

	Load Cases	Nodes	Loads		
			P_x (kips)	P_y (kips)	P_z (kips)
Case 1	1	1	1.0	-10.0	-10.0
		2	0.0	-10.0	-10.0
		3	0.5	0.0	0.0
		6	0.6	0.0	0.0

Table 18 The load case 2 and case 3 for the 25-bar spatial truss structure

	Load Cases	Nodes	Loads		
			P_x (kips)	P_y (kips)	P_z (kips)
Case 2 & 3	2	1	0.0	20.0	-5.0
		2	0.0	-20.0	-5.0
	3	1	1.0	10.0	-5.0
		2	0.0	10.0	-5.0
		3	0.5	0.0	0.0
		6	0.5	0.0	0.0

Table 19 The available cross-section areas of the ASIC code

No.	in^2	mm^2	No.	in^2	mm^2
1	0.111	71.613	33	3.840	2477.414
2	0.141	90.968	34	3.870	2496.769
3	0.196	126.451	35	3.880	2503.221
4	0.250	161.290	36	4.180	2696.769
5	0.307	198.064	37	4.220	2722.575
6	0.391	252.258	38	4.490	2896.768
7	0.442	285.161	39	4.590	2961.284
8	0.563	363.225	40	4.800	3096.768
9	0.602	388.386	41	4.970	3206.445
10	0.766	494.193	42	5.120	3303.219
11	0.785	506.451	43	5.740	3703.218
12	0.994	641.289	44	7.220	4658.055
13	1.000	645.160	45	7.970	5141.925
14	1.228	792.256	46	8.530	5503.215
15	1.266	816.773	47	9.300	5999.988
16	1.457	939.998	48	10.850	6999.986
17	1.563	1008.385	49	11.500	7419.340
18	1.620	1045.159	50	13.500	8709.660
19	1.800	1161.288	51	13.900	8967.724
20	1.990	1283.868	52	14.200	9161.272
21	2.130	1374.191	53	15.500	9999.980
22	2.380	1535.481	54	16.000	10322.560
23	2.620	1690.319	55	16.900	10903.204
24	2.630	1696.771	56	18.800	12129.008
25	2.880	1858.061	57	19.900	12838.684
26	2.930	1890.319	58	22.000	14193.520
27	3.090	1993.544	59	22.900	14774.164
28	1.130	729.031	60	24.500	15806.420
29	3.380	2180.641	61	26.500	17096.740
30	3.470	2238.705	62	28.000	18064.480
31	3.550	2290.318	63	30.000	19354.800
32	3.630	2341.931	64	33.500	21612.860

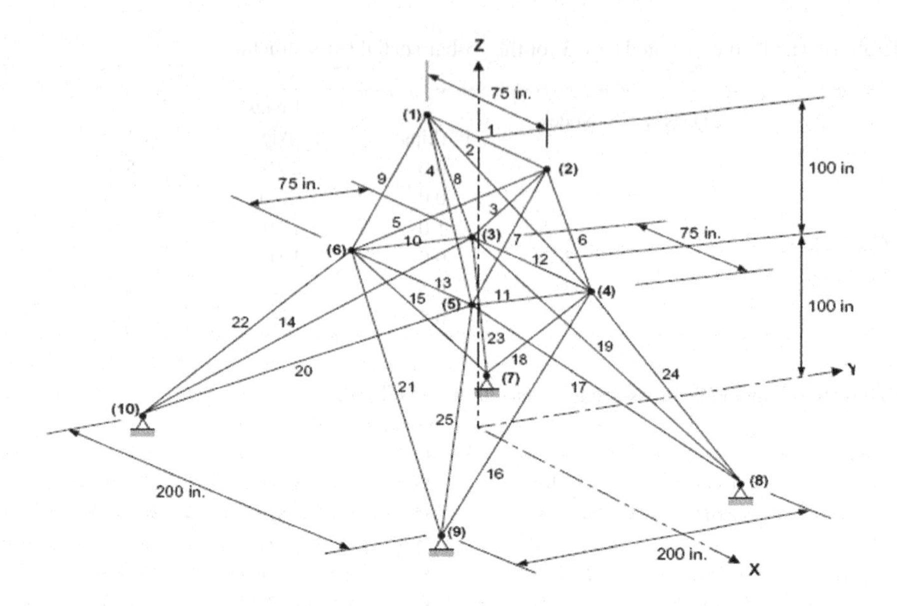

Fig. 19 A 25-bar spatial truss structure

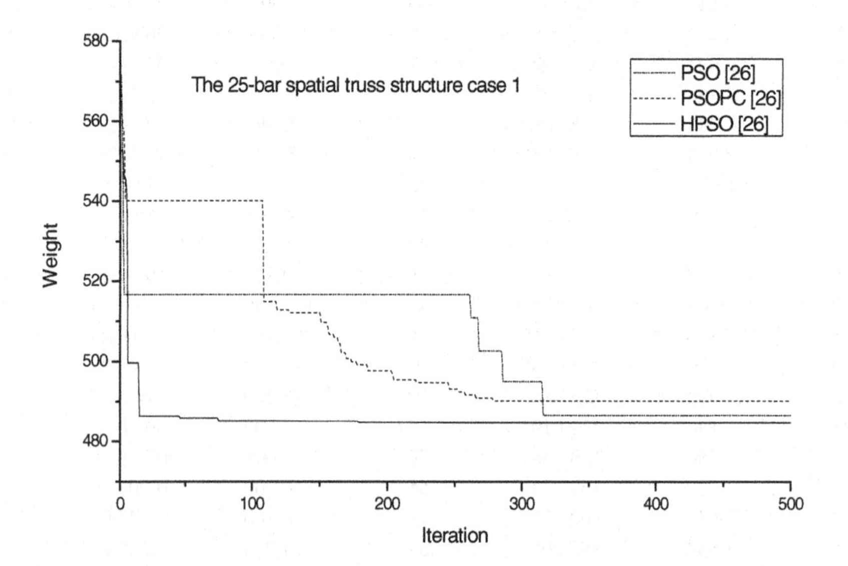

Fig. 20 Comparison of convergence rates for the 25-bar spatial truss structure (case 1)

implemented. Case 1: The discrete variables are selected from the set D= {0.1, 0.2, 0.3, 0.4, 0.5, 0.6, 0.7, 0.8, 0.9, 1.0, 1.1, 1.2, 1.3, 1.4, 1.5, 1.6, 1.7, 1.8, 1.9, 2.0, 2.1, 2.2, 2.3, 2.4, 2.6, 2.8, 3.0, 3.2, 3.4} (in^2). The loads are shown in Table 17; Case 2: The discrete variables are selected from the set D= {0.01, 0.4, 0.8, 1.2,

1.6, 2.0, 2.4, 2.8, 3.2, 3.6, 4.0, 4.4, 4.8, 5.2, 5.6, 6.0} (in^2). The loads are shown in Table 18. Case 3: The discrete variables are selected from the American Institute of Steel Construction (AISC) Code, which is shown in Table 19. The loads are shown in Table 18. A maximum number of 500 iterations is imposed for three cases.

Table 20, Table 21 and Table 22 show the comparison of optimal design results for the 25-bar spatial truss structure under three load cases while Fig. 20, Fig.21 and Fig.22 show comparison of convergence rates. For all load cases of this structure, three algorithms can achieve the optimal solution after 500 iterations. But Figures 20, 21 and 22 show that the HPSO algorithm has the fastest convergence rate.

Table 20 Comparison of optimal designs for the 25-bar spatial truss structure (case 1)

Variables (in^2)	Case 1					
	Wu [24]	Rajeev [28]	Lee [13]	Li [26] PSO	Li [26] PSOPC	Li [26] HPSO
A_1	0.1	0.1	0.1	0.4	0.1	0.1
A_2~A_5	0.5	1.8	0.3	0.6	1.1	0.3
A_6~A_9	3.4	2.3	3.4	3.5	3.1	3.4
A_{10}~A_{11}	0.1	0.2	0.1	0.1	0.1	0.1
A_{12}~A_{13}	1.5	0.1	2.1	1.7	2.1	2.1
A_{14}~A_{17}	0.9	0.8	1.0	1.0	1.0	1.0
A_{18}~A_{21}	0.6	1.8	0.5	0.3	0.1	0.5
A_{22}~A_{25}	3.4	3.0	3.4	3.4	3.5	3.4
Weight (lb)	486.29	546.01	484.85	486.54	490.16	484.85

Table 21 Comparison of optimal designs for the 25-bar spatial truss structure (case 2)

Variables (in^2)	Case 2					
	Wu [24]	Ringertz [29]	Lee [13]	Li [26] PSO	Li [26] PSOPC	Li [26] HPSO
A_1	0.4	0.01	0.01	0.01	0.01	0.01
A_2~A_5	2.0	1.6	2.0	2.0	2.0	2.0
A_6~A_9	3.6	3.6	3.6	3.6	3.6	3.6
A_{10}~A_{11}	0.01	0.01	0.01	0.01	0.01	0.01
A_{12}~A_{13}	0.01	0.01	0.01	0.4	0.01	0.01
A_{14}~A_{17}	0.8	0.8	0.8	0.8	0.8	0.8
A_{18}~A_{21}	2.0	2.0	1.6	1.6	1.6	1.6
A_{22}~A_{25}	2.4	2.4	2.4	2.4	2.4	2.4
Weight (lb)	563.52	568.69	560.59	566.44	560.59	560.59

Table 22 Comparison of optimal designs for the 25-bar spatial truss structure (case 3)

Variables (in^2)	Case 3			
	Wu [24]	Li [26] PSO	Li [26] PSOPC	Li [26] HPSO
A_1	0.307	1.0	0.111	0.111
A_2~A_5	1.990	2.62	1.563	2.130
A_6~A_9	3.130	2.62	3.380	2.880
A_{10}~A_{11}	0.111	0.25	0.111	0.111
A_{12}~A_{13}	0.141	0.307	0.111	0.111
A_{14}~A_{17}	0.766	0.602	0.766	0.766
A_{18}~A_{21}	1.620	1.457	1.990	1.620
A_{22}~A_{25}	2.620	2.880	2.380	2.620
Weight (lb)	556.43	567.49	556.90	551.14

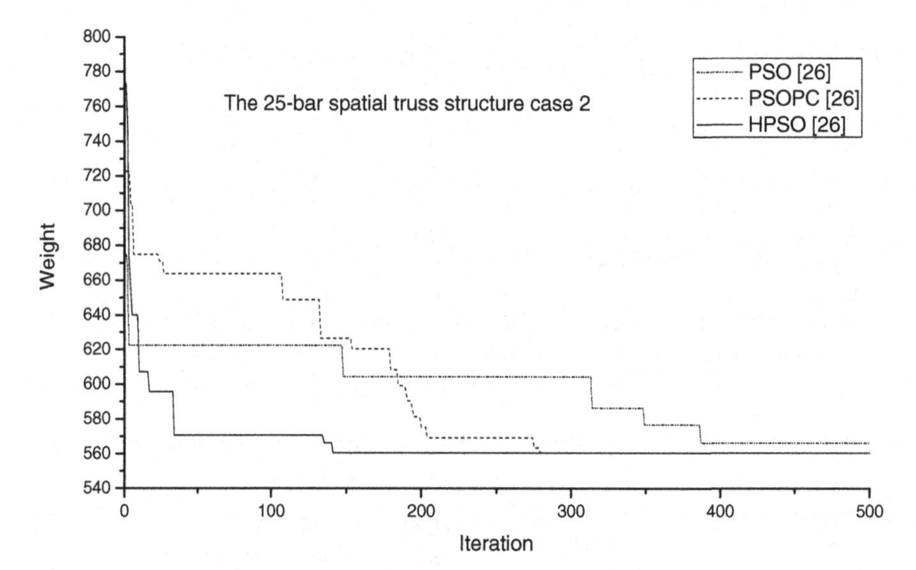

Fig. 21 Comparison of convergence rates for the 25-bar spatial truss structure (case 2)

5.3.4 A 52-Bar Planar Truss Structure

A 52-bar planar truss structure, shown in Fig. 23, has been analysed by Wu [24] Lee [13] and Li [26]. The members of this structure are divided into 12 groups: (1) A_1~A_4, (2) A_5~A_6, (3) A_7~A_8, (4) A_9~A_{10}, (5) A_{11}~A_{14}, (6) A_{15}~A_{18}, and (7) A_{19}~A_{22}. The material density is 7860.0 kg/m^3 and the modulus of elasticity is 2.07× 10^{5}MPa. The members are subjected to stress limitations of ±180MPa. Both of the loads, P_x =100kN, P_y =200kN are considered. The discrete variables are selected from the Table 19. A maximum number of 3,000 iterations is imposed.

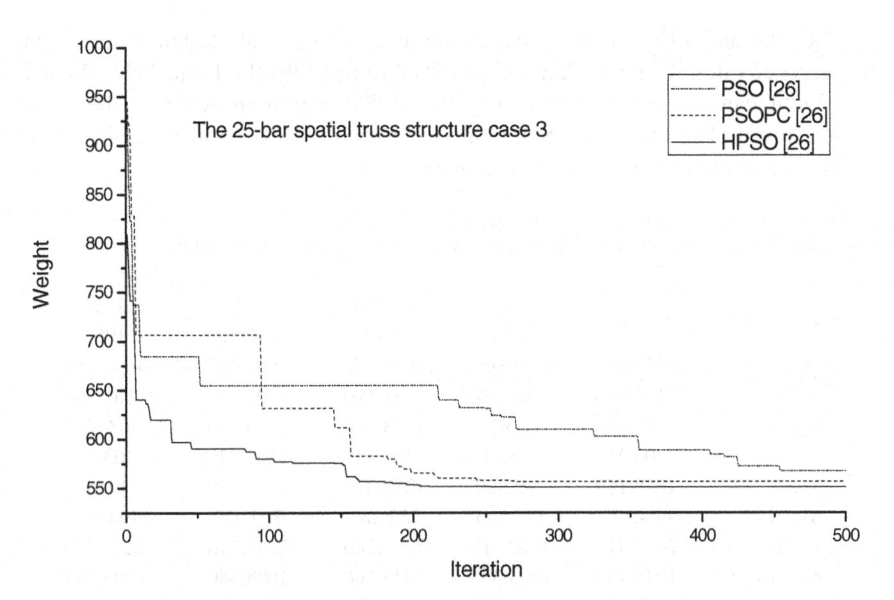

Fig. 22 Comparison of convergence rates for the 25-bar spatial truss structure (case 3)

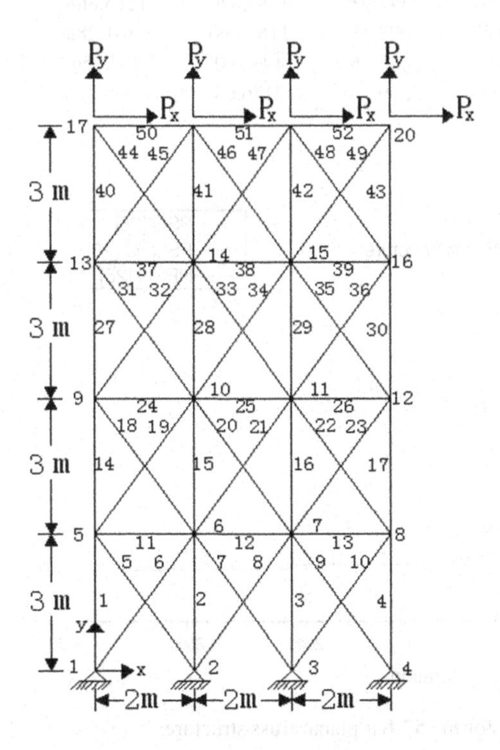

Fig. 23 A 52-bar planar truss structure

Table 23 and Fig. 24 give the comparison of optimal design results and convergence rates of 52-bar planar truss structure respectively. From Table 23 and Fig. 24, it can be observed that only the HPSO algorithm achieves the good optimal result. The PSO and PSOPC algorithms do not get optimal results when the maximum number of iterations is reached.

Table 23 Comparison of optimal designs for the 52-bar planar truss structure

Variables (mm^2)	Wu [24]	Lee [13]	Li [26] PSO	Li [26] PSOPC	Li [26] HPSO
A_1~A_4	4658.055	4658.055	4658.055	5999.988	4658.055
A_5~A_{10}	1161.288	1161.288	1374.190	1008.380	1161.288
A_{11}~A_{13}	645.160	506.451	1858.060	2696.770	363.225
A_{14}~A_{17}	3303.219	3303.219	3206.440	3206.440	3303.219
A_{18}~A_{23}	1045.159	940.000	1283.870	1161.290	940.000
A_{24}~A_{26}	494.193	494.193	252.260	729.030	494.193
A_{27}~A_{30}	2477.414	2290.318	3303.220	2238.710	2238.705
A_{31}~A_{36}	1045.159	1008.385	1045.160	1008.380	1008.385
A_{37}~A_{39}	285.161	2290.318	126.450	494.190	388.386
A_{40}~A_{43}	1696.771	1535.481	2341.93	1283.870	1283.868
A_{44}~A_{49}	1045.159	1045.159	1008.38	1161.290	1161.288
A_{50}~A_{52}	641.289	506.451	1045.16	494.190	792.256
Weight (kg)	1970.142	1906.76	2230.16	2146.63	1905.495

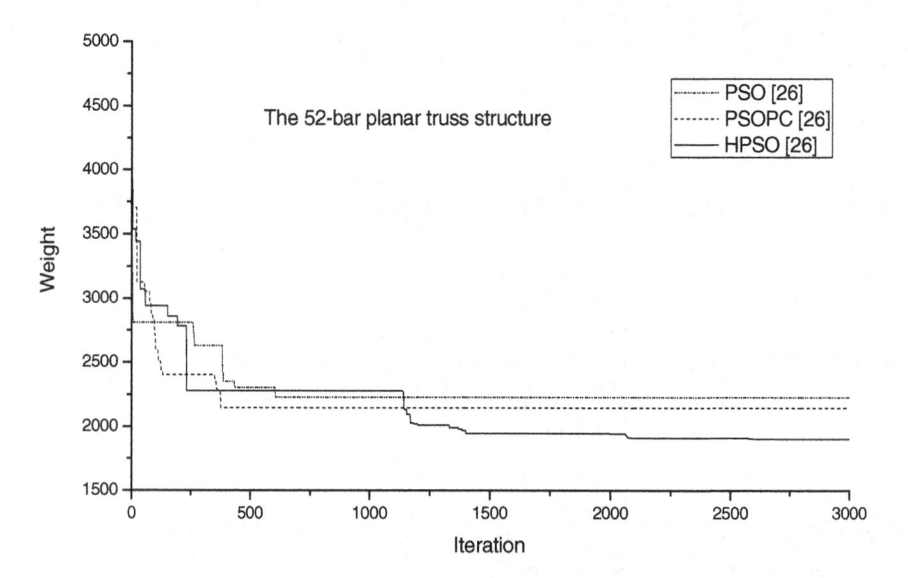

Fig. 24 Comparison of convergence rates for the 52-bar planar truss structure

5.3.5 A 72-Bar Spatial Truss Structure

A 72-bar spatial truss structure, shown in Fig. 25, has been studied by Wu [24] Lee [13] and Li [26]. The material density is 0.1 lb/in.^{3} and the modulus of elasticity is 10,000 ksi. The members are subjected to stress limitations of ±25 ksi. The uppermost nodes are subjected to displacement limitations of ±0.25 in. both in *x* and *y* directions. Two load cases are listed in Table 24. There are 72 members, which are divided into 16 groups, as follows: (1) A_1~A_4, (2) A_5~$A_{12,}$ (3) A_{13}~A_{16}, (4) A_{17}~A_{18}, (5) A_{19}~A_{22}, (6) A_{23}~A_{30} (7) A_{31}~A_{34}, (8) A_{35}~A_{36}, (9) A_{37}~A_{40}, (10) A_{41}~$A_{48,}$ (11) A_{49}~A_{52}, (12) A_{53}~A_{54}, (13) A_{55}~A_{58}, (14) A_{59}~A_{66} (15) A_{67}~A_{70}, (16) A_{71}~A_{72}. There are two optimization cases to be implemented. Case 1: The discrete variables are selected from the set D={0.1, 0.2, 0.3, 0.4, 0.5, 0.6, 0.7, 0.8, 0.9, 1.0, 1.1, 1.2, 1.3, 1.4, 1.5, 1.6, 1.7, 1.8, 1.9, 2.0, 2.1, 2.2, 2.3, 2.4, 2.5, 2.6, 2.7,

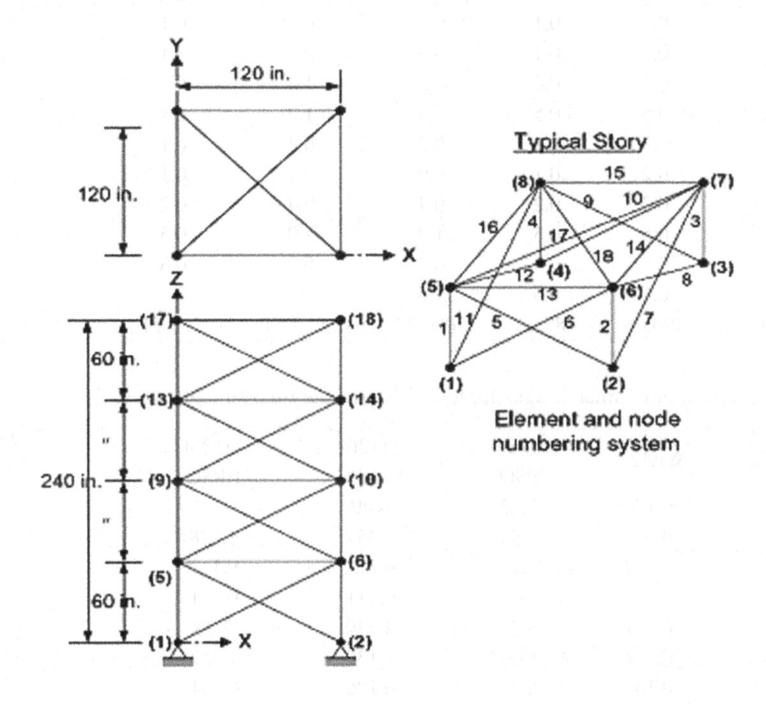

Fig. 25 The 72-bar spatial truss structure

Table 24 The load cases for the 72-bar spatial truss structure

Nodes	Load Case 1			Load Case 2		
	P_X (kips)	P_Y (kips)	P_Z (kips)	P_X (kips)	P_Y (kips)	P_Z (kips)
17	5.0	5.0	-5.0	0.0	0.0	-5.0
18	0.0	0.0	0.0	0.0	0.0	-5.0
19	0.0	0.0	0.0	0.0	0.0	-5.0
20	0.0	0.0	0.0	0.0	0.0	-5.0

2.8, 2.9, 3.0, 3.1, 3.2} (in^2); Case 2: The discrete variables are selected from the Table 19. A maximum number of 1,000 iterations is imposed.

Table 25 and Table 26 are the comparison of optimal design results.

Table 25 Comparison of optimal designs for the 72-bar spatial truss structure (case 1)

Variables (in^2)	Wu [24]	Lee [13]	Li [26] PSO	Li [26] PSOPC	Li [26] HPSO
A_1~A_4	1.5	1.9	2.6	3.0	2.1
A_5~A_{12}	0.7	0.5	1.5	1.4	0.6
A_{13}~A_{16}	0.1	0.1	0.3	0.2	0.1
A_{17}~A_{18}	0.1	0.1	0.1	0.1	0.1
A_{19}~A_{22}	1.3	1.4	2.1	2.7	1.4
A_{23}~A_{30}	0.5	0.6	1.5	1.9	0.5
A_{31}~A_{34}	0.2	0.1	0.6	0.7	0.1
A_{35}~A_{36}	0.1	0.1	0.3	0.8	0.1
A_{37}~A_{40}	0.5	0.6	2.2	1.4	0.5
A_{41}~A_{48}	0.5	0.5	1.9	1.2	0.5
A_{49}~A_{52}	0.1	0.1	0.2	0.8	0.1
A_{53}~A_{54}	0.2	0.1	0.9	0.1	0.1
A_{55}~A_{58}	0.2	0.2	0.4	0.4	0.2
A_{59}~A_{66}	0.5	0.5	1.9	1.9	0.5
A_{67}~A_{70}	0.5	0.4	0.7	0.9	0.3
A_{71}~A_{72}	0.7	0.6	1.6	1.3	0.7
Weight (lb)	400.66	387.94	1089.88	1069.79	388.94

Table 26 Comparison of optimal designs for the 72-bar spatial truss structure (case 2)

Variables (in^2)	Wu [24]	Li [26] PSO	Li [26] PSOPC	Li [26] HPSO
A_1~A_4	0.196	7.22	4.49	4.97
A_5~A_{12}	0.602	1.80	1.457	1.228
A_{13}~A_{16}	0.307	1.13	0.111	0.111
A_{17}~A_{18}	0.766	0.196	0.111	0.111
A_{19}~A_{22}	0.391	3.09	2.620	2.88
A_{23}~A_{30}	0.391	0.785	1.130	1.457
A_{31}~A_{34}	0.141	0.563	0.196	0.141
A_{35}~A_{36}	0.111	0.785	0.111	0.111
A_{37}~A_{40}	1.800	3.09	1.266	1.563
A_{41}~A_{48}	0.602	1.228	1.457	1.228
A_{49}~A_{52}	0.141	0.111	0.111	0.111
A_{53}~A_{54}	0.307	0.563	0.111	0.196
A_{55}~A_{58}	1.563	1.990	0.442	0.391
A_{59}~A_{66}	0.766	1.620	1.457	1.457
A_{67}~A_{70}	0.141	1.563	1.228	0.766
A_{71}~A_{72}	0.111	1.266	1.457	1.563
Weight (lb)	427.203	1209.48	941.82	933.09

The Fig.26 and Fig.27 are comparison of convergence rates for the 72-bar spatial truss structure in two load cases.

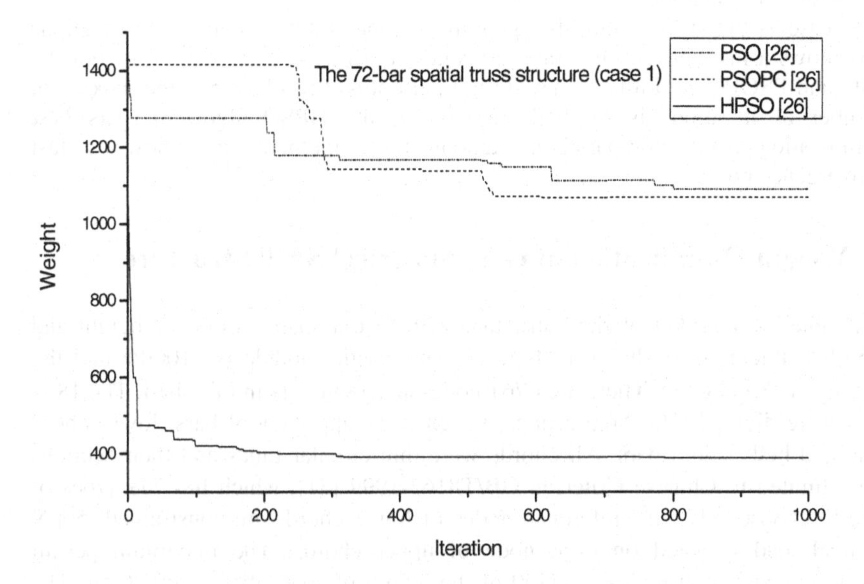

Fig. 26 Comparison of convergence rates for the 72-bar spatial truss structure (case 1)

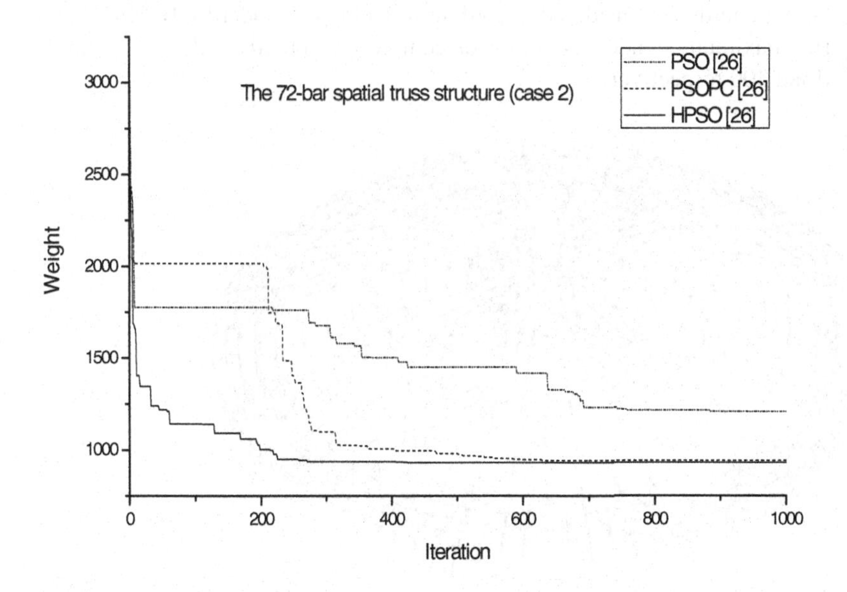

Fig. 27 Comparison of convergence rates for the 72-bar spatial truss structure (case 2)

For both of the cases, it seems that Wu's results [24] achieve smaller weight. However, we discovered that both of these results do not satisfy the constraints. The results are unacceptable.

In case 1, the HPSO algorithm gets the optimal solution after 1000 iterations and shows a fast convergence rate, especially during the early iterations. For the PSO and PSOPC algorithms, they do not get optimal results when the maximum number of iterations is reached. In case 2, the HPSO algorithm gets best optimization result comparatively among three methods and shows a fast convergence rate.

6 Weight Optimization of Grid Spherical Shell Structure

A double-layer grid steel shell structure with 83.6m span, 14.0m arc height and 1.5 shell thickness is shown in Fig. 28. The elastic module is 210GPa and the density is 7850 kg/m^3. There are 6761 nodes and 1834 bars in this shell. The 1834 bars were divided into three groups, which were upper chord bars, lower chord bars and belly chord bars. All chords were thin circular tubes and their sections were limited to Chinese Criterion GB/T8162-1999 [31], which has 779 types of size to choose. The circumference nodes of lower chords are constrained. 50kN vertical load is acted on each node of upper chords. The maximum permit displacement for all nodes is 1/400 of the length of span, that is ±0.209m. The maximum permit stress for all chord bars is ±215MPa. The stability of compressive chords is considered according to Chinese Standard GB50017-2003 [32]. The maximum slenderness ratio for compressive chords and tensile chords are 180 and 300 respectively.

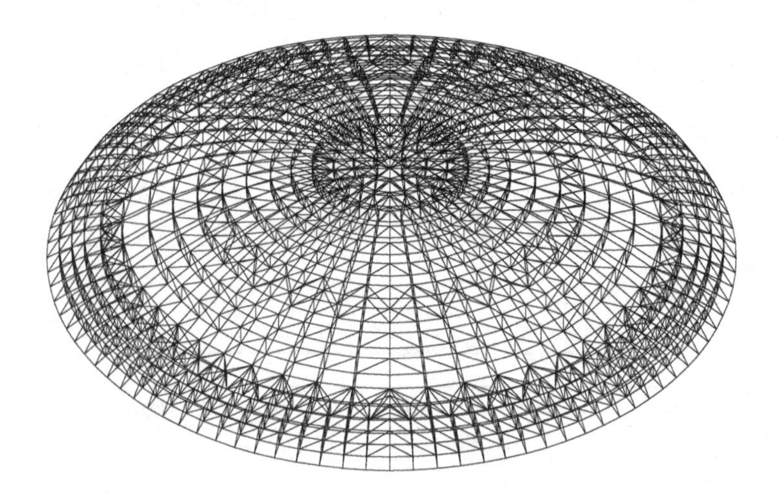

Fig. 28 The double layer reticulated spherical shell structure

The optimization results [26] are shown in Table 27. The convergence velocity is shown in Figure 29. It can be seen from Fig. 29 that HPSO can be used effectively to optimize the complicated engineering structures.

Table 27 The optimal solution for the double layer reticulated spherical shell structure

Upper chord bars	Lower chord bars	Belly chord bars	Weight (kg)
φ108×4	φ83×3.5	φ89×3.5	148811.71

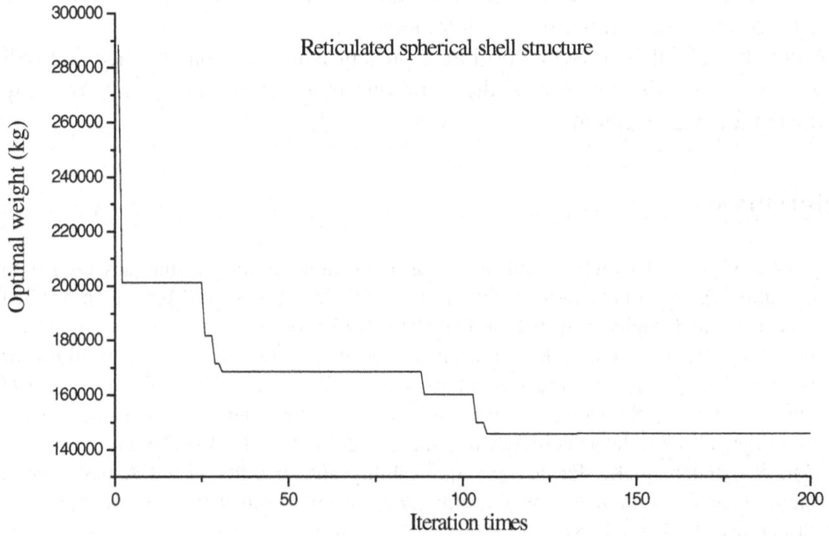

Fig. 29 The convergence rate of the HPSO for the double layer grid spherical shell structure

As there are only three group optimal variables chosen in this example, the convergence rate is considerably fast, within only about 120 iterations. Anyway, the weight optimization for a reticulated shell structure with 1834 bars is a very complicated engineering problem. It is almost impossible to get an optimal solution using traditional optimal methods while the HPSO has an ability of handling complex structural optimization problems effectively.

7 Conclusions Remarks

In this chapter, a harmony particle swarm optimizer (HPSO), based on the particle swarm optimizer with passive congregation (PSOPC) and the harmony search (HS) algorithm, was presented. The HPSO algorithm handles the constraints of variables using the harmony search scheme in corporation with the 'fly-back mechanism' method used to deal with the problem-specific constraints. Compared with the PSO and the PSOPC algorithms, the HPSO algorithm does not allow any particles to fly outside the boundary of the variables and makes a full use of algorithm flying behaviour of each particle. Thus this algorithm performs more efficient than the others.

The efficiency of the HPSO algorithm presented was tested for optimum design problems of planar and spatial pin-connected structures with continuous and discrete variables. A double-layer grid shell structure was also used to test the HPSO. All the results show that the HPSO algorithm has better search behaviour avoiding premature convergence while rapidly converging to the optimal solution. And the HPSO algorithm converges more quickly than the PSO and the PSOPC algorithms, in particular, in the early iterations.

A drawback of this HPSO algorithm at present is that its convergence rate will slow down when the number of the iterations increase. Further study is being conducted for improvement.

References

1. Coello, C.A.C.: Theoretical and numerical constraint-handling techniques used with evolutionary algorithms: a survey of the state of the art. Computer Methods in Applied Mechanics and Engineering 191(11-12), 1245–1287 (2002)
2. Nanakorn, P., Meesomklin, K.: An adaptive penalty function in genetic algorithms for structural design optimization. Computers and Structures 79(29-30), 2527–2539 (2001)
3. Deb, K., Gulati, S.: Design of truss-structures for minimum weight using genetic algorithms. Finite Elements in Analysis and Design 37(5), 447–465 (2001)
4. Ali, N., Behdinan, K., Fawaz, Z.: Applicability and viability of a GA based finite element analysis architecture for structural design optimization. Computers and Structures 81(22-23), 2259–2271 (2003)
5. Kennedy, J., Eberhart, R.: Particle swarm optimization. In: IEEE International Conference on Neural Networks, vol. 4, pp. 1942–1948. IEEE Computer Society Press, Los Alamitos (1995)
6. Kennedy, J., Eberhart, R.C.: Swarm intelligence. Morgan Kaufmann Publishers, San Francisco (2001)
7. Geem, Z.W.: Particle-swarm harmony search for water network design. Engineering Optimization 41(4), 297–311 (2009)
8. He, S., Wu, Q.H., Wen, J.Y., Saunders, J.R., Paton, R.C.: A particle swarm optimizer with passive congregation. BioSystem 78(1-3), 135–147 (2004)
9. Prempain, H.S., Qh, W.: An improved particle swarm optimizer for mechanical design optimization problems. Engineering Optimization 36(5), 585–605 (2004)
10. Davis, L.: Genetic algorithms and simulated annealing. Pitman, London (1987)
11. Le Riche, R.G, Knopf-Lenoir, C., Haftka, R.T.: A segregated genetic algorithm for constrained structural optimization. In: Sixth International Conference on Genetic Algorithms, pp. 558–565. University of Pittsburgh, Morgan Kaufmann, San Francisco (1995)
12. Li, L.J., Ren, F.M., Liu, F., Wu, Q.H.: An improved particle swarm optimization method and its application in civil engineering. In: Proceedings of the Fifth International Conference on Engineering Computational Technology, Spain, paper 42 (2006)

13. Lee, K.S., Geem, Z.W., Lee, S.H., Bae, K.W.: The harmony search heuristic algorithm for discrete structural optimization. Engineering Optimization 37(7), 663–684 (2005)
14. Shi, Y., Eberhart, R.C.: A modified particle swarm optimizer. In: Proc. IEEE Inc. Conf. On Evolutionary Computation, pp. 303–308 (1997)
15. Geem, Z.W., Kim, J.H., Loganathan, G.V.: A new heuristic optimization algorithm: harmony search. Simulation 76(2), 60–68 (2001)
16. Lee, K.S., Geem, Z.W.: A new structural optimization method based on the harmony search algorithm. Computers and Structures 82(9-10), 781–798 (2004)
17. Schmit Jr., L.A., Farshi, B.: Some approximation concepts for structural synthesis. AIAA J. 12(5), 692–699 (1974)
18. Rizzi, P.: Optimization of multiconstrained structures based on optimality criteria. In: AIAA/ASME/SAE 17th Structures, Structural Dynamics and Materials Conference, King of Prussia, PA (1976)
19. Li, L.J., Huang, Z.B., Liu, F.: A heuristic particle swarm optimizer (HPSO) for optimization of pin connected structures. Computers and Structures 85(7-8), 340–349 (2007)
20. Khot, N.S., Berke, L.: Structural optimization using optimality criteria methods. In: Atrek, E., Gallagher, R.H., Ragsdell, K.M., Zienkiewicz, O.C. (eds.) New directions in optimum structural design. John Wiley, New York (1984)
21. Adeli, H., Kumar, S.: Distributed genetic algorithm for structural optimization. J. Aerospace Eng., ASCE 8(3), 156–163 (1995)
22. Sarma, K.C., Adeli, H.: Fuzzy genetic algorithm for optimization of steel structures. J. Struct. Eng., ASCE 126(5), 596–604 (2000)
23. He, D.K., Wang, F.L., Mao, Z.Z.: Study on application of genetic algorithm in discrete variables optimization. Journal of System Simulation 18(5), 1154–1156 (2006)
24. Wu, S.J., Chow, P.T.: Steady-state genetic algorithms for discrete optimization of trusses. Computers and Structures 56(6), 979–991 (1995)
25. Parsopoulos, K.E., Vrahatis, M.N.: Recent approaches to global optimization problems through particle swarm optimization. Natural Computing 12(1), 235–306 (2002)
26. Li, L.J., Huang, Z.B., Liu, F.: A heuristic particle swarm optimization method for truss structures with discrete variables. Computers and Structures 87(7-8), 435–443 (2009)
27. Li, L.J., Huang, Z.B., Liu, F.: An improved particle swarm optimizer for truss structure optimization. In: Wang, Y., Cheung, Y.-m., Liu, H. (eds.) CIS 2006. LNCS, vol. 4456, pp. 1–10. Springer, Heidelberg (2007)
28. Rajeev, S., Krishnamoorthy, C.S.: Discrete optimization of structures using genetic algorithm. Journal of Structural Engineering, ASCE 118(5), 1123–1250 (1992)
29. Ringertz, U.T.: On methods for discrete structural constraints. Engineering Optimization 13(1), 47–64 (1988)
30. Zhang, Y.N., Liu, J.P., Liu, B., Zhu, C.Y., Li, Y.: Application of improved hybrid genetic algorithm to optimize. Journal of South China University of Technology 33(3), 69–72 (2003)
31. Chinese Criterion, Seamless steel tubes for structural purposes, GB/T8162-1999. Standards Press of China, Beijing (1999)
32. Chinese Standard, Code for design of steel structures, GB50017-2003. China Architecture and Building Press, Beijing (2006)

Hybrid Algorithm of Harmony Search, Particle Swarm and Ant Colony for Structural Design Optimization

A. Kaveh[1] and S. Talatahari[2]

Abstract. This chapter considers the implementation of the heuristic particle swarm ant colony optimization (HPSACO) methodology to find an optimum design of different types of structures. HPSACO is an efficient hybridized approach based on the harmony search scheme, particle swarm optimizer, and ant colony optimization. HPSACO utilizes a particle swarm optimization with a passive congregation algorithm as a global search, and the idea of ant colony approach worked as a local search. The harmony search-based mechanism is used to handle the variable constraints. In the discrete HPSACO, agents are allowed to select discrete values from the permissible list of cross sections. The efficiency of the HPSACO algorithm is investigated to find an optimum design of truss structures with continuous or discrete search domains and for frame structures with a discrete search domain. The results indicate that the HPSACO is a quite effective algorithm to find the optimum solution of structural optimization problems with continuous or discrete variables.

1 Introduction

Structural design optimization is a critical and challenging activity that has received considerable attention in the last two decades [1]. A high number of design variables, largeness of the search space and controlling a great number of design constraints are major preventive factors in performing optimum design in a reasonable time. Despite these facts, designers and owners have always desired to have optimal structures [2]. Therefore, different methods of structural optimization have been introduced which can be categorized in two general groups: classical methods and heuristic approaches.

Classical optimization methods are often based on mathematical programming. Many of these methods require substantial gradient information, and final results depend on the initially selected points. The number of computational operations increases as the design variables of a structure becomes greater and the solution

---

[1] Centre of Excellence for Fundamental Studies in Structural Engineering, Iran University of Science and Technology, Narmak, Tehran-16, Iran
Email: alikaveh@iust.ac.ir

[2] Department of Civil Engineering, University of Tabriz, Tabriz, Iran
Email: siamak.talat@gmail.com

Z.W. Geem (Ed.): Harmony Search Algo. for Structural Design Optimization, SCI 239, pp. 159–198.
springerlink.com

does not necessarily correspond to the global optimum or even the neighborhood of it, in some cases.

The computational drawbacks of classical numerical methods have forced researchers to rely on heuristic algorithms such as genetic algorithms (GAs), particle swarm optimizer (PSO), ant colony optimization (ACO) and harmony search (HS). These methods have attracted a great deal of attention, because of their high potential for modeling engineering problems in environments which have been resistant to a solution by classic techniques. They do not require gradient information and possess better global search abilities than the conventional optimization algorithms. Although these are approximate methods (i.e. their solutions are good, but not provably optimal), they do not require the derivatives of the objective function and constraints [3]. Having in common the processes of natural evolution, these algorithms share many similarities: each maintains a population of solutions which are evolved through random alterations and selection. The differences between these procedures lie in the representation technique utilized to encode the candidates, the type of alterations used to create new solutions, and the mechanism employed for selecting new patterns.

The genetic algorithm is one of the heuristic algorithms initially suggested by Holland, and developed and extended by some of his students, Goldberg and De Jong. These algorithms simulate a natural genetics mechanism for synthetic systems based on operators that are duplicates of natural ones. In the last decade, GA has been used in the optimum structural design. One of the first applications was the weight minimization of a 10-bar truss by Goldberg and Samtani [4]. Also, many researchers have used genetic search in the design of various structures in which the search space was non-convex or discrete, Hajela [5], Rajeev and Krishnamoorthy [6,7], Koumousis and Georgious [8], Hajela and Lee [9], Wu and Chow [10], Soh and Yang [11], Camp *et al.* [12], Shrestha and Ghaboussi [13], Pezeshek *et al.* [14] Erbatur *et al.* [15], Coello and Christiansen [16], Greiner *et al.* [17], Kameshki and Saka [18-20], Saka [21, 22], and Kaveh and colleagues [23-28], among many others.

Application of swarm intelligence for optimization was first suggested by Eberhart and Kennedy [29] under the name of particle swarm optimization (PSO). The strength of PSO is underpinned by the fact that decentralized biological creatures can often accomplish complex goals by cooperation. A standard PSO algorithm is initialized with a population (swarm) of random potential solutions (particles). Each particle iteratively moves across the search space and is attracted to the position of the best fitness historically achieved by the particle itself (local best) and by the best among the neighbors of the particle (global best) [30]. Compared to other evolutionary algorithms based on heuristics, the advantages of PSO consist of easy implementation and a smaller number of parameters to be adjusted. Therefore, it has been widely employed for structural optimization problems [31-35]. However, it is known that the PSO algorithm had difficulties in controlling the balance between exploration (global investigation of the search place) and exploitation (the fine search around a local optimum) [36].

Ant colony optimization (ACO) was first proposed by Dorigo [37, 38] as a multi-agent approach to solve difficult combinatorial optimization problems and it has been applied to various engineering problems in recent years [39-44]. ACO was inspired by the observation of real ant colonies. Ants are social insects whose behavior is directed more to the survival of the colony as a whole than to that of a single individual component of the colony. An important behavior of ant colonies is their foraging behavior, and in particular, how the ants can find shortest paths between food sources and their nest. While walking from food sources to the nest and vice versa, ants deposit on the ground a substance called pheromone. Ants can smell pheromone and when choosing their way, they tend to choose, in probability, paths marked by strong pheromone concentrations. When more paths are available from the nest to a food source, a colony of ants will be able to exploit the pheromone trails left by the individual ants to discover the shortest path from the nest to the food source and back. One basic idea of the ACO approach is to employ the counterpart of the pheromone trail used by real ants as an indirect communication and as a form of memory of previously found solutions.

The harmony search method, as discussed in the previous chapters, is another robust heuristic optimization technique that imitates the musical performance process which takes place when a musician searches for a better state of harmony. Jazz improvisation seeks to find musically pleasing harmony similar to the optimum design process which seeks to find the optimum solution. The pitch of each musical instrument determines the aesthetic quality, just as the objective function value is determined by the set of values assigned to each decision variable. This approach is suggested by Geem *et al.* in 2001 [45] and first applied to a design of water distribution network. Since then, the algorithm has attracted many researchers due to its simplicity and effectiveness [1, 46-51].

Although there are several papers utilizing heuristic methods in the structural optimization field, using an individual heuristic method has often had some drawbacks because usually each method is suitable for solving only a specific group of problems. Preference for a special method will differ depending on the kind of the problem being studied. One technique to overcome these problems is hybridizing various methods to reach a robust approach.

In this chapter, the implementation of an efficient hybrid algorithm based on harmony search, particle swarm and ant colony strategies, namely heuristic particle swarm ant colony optimization (HPSACO), is developed to find an optimum design of truss structures with continuous or discrete domains and to find frame structures with a discrete search domain.

2 Review of PSO, ACO and HS Algorithms

Since HPSACO methodology is based on PSO, ACO and HS, in order to make the chapter self-explanatory, the characteristics of these algorithms are briefly explained in this section.

2.1 Particle Swarm Optimization

Particle swarm optimization (PSO) is a stochastic optimization method capable of handling non differentiable, nonlinear, and multi module objective functions. The

PSO method is motivated from the social behavior of bird flocking and fish schooling [29]. PSO has a population of individuals that move through search space and each individual has a velocity that acts as an operator to obtain a new set of individuals. Individuals, called particles, adjust their movements depending on both their own experience and the population's experience. Effectively, each particle continuously focuses and refocuses on the effort of its search according to both the local and global best. This behavior mimics the cultural adaptation of a biological agent in a swarm: it evaluates its own position based on certain fitness criteria, compares it to others, and imitates the best position in the entire swarm [30].

Through the updating process, each particle moves by adding a change velocity $\mathbf{V}_i^{k+1}$ to the current position $\mathbf{X}_i^k$ as follows

$$\mathbf{X}_i^{k+1} = \mathbf{X}_i^k + \mathbf{V}_i^{k+1} \tag{1}$$

The velocity is a combination of three contributing factors:

1. Previous velocity, $\mathbf{V}_i^k$, considering former attempts;
2. Movement in the direction of the local best, $\mathbf{P}_i^k$, using the autobiographical memory;
3. Movement in the direction of the global best, $\mathbf{P}_g^k$, based on the publicized knowledge.

The mathematical relationship can be expressed as

$$\mathbf{V}_i^{k+1} = \omega\mathbf{V}_i^k + c_1 r_1(\mathbf{P}_i^k - \mathbf{X}_i^k) + c_2 r_2(\mathbf{P}_g^k - \mathbf{X}_i^k) \tag{2}$$

where ω is an inertia weight to control the influence of the previous velocity; r_1 and r_2 are two random numbers uniformly distributed in the range of (0, 1); c_1 and c_2 are two acceleration constants. $\mathbf{P}_i^k$ is the best position of the i th particle up to iteration k and $\mathbf{P}_g^k$ is the best position among all particles in the swarm up to iteration k . $\mathbf{P}_i^k$ and $\mathbf{P}_g^k$ are given by the following equations

$$\mathbf{P}_i^k = \begin{cases} \mathbf{P}_i^{k-1} & f(\mathbf{X}_i^k) \geq f(\mathbf{P}_i^{k-1}) \\ \mathbf{X}_i^k & f(\mathbf{X}_i^k) < f(\mathbf{P}_i^{k-1}) \end{cases} \tag{3}$$

$$\mathbf{P}_g^k = \{\mathbf{P}_i^k \mid f(\mathbf{P}_i^k) = \min(f(\mathbf{P}_g^{k-1}) \wedge f(\mathbf{P}_j^k), j = 1,2,.., M)\} \tag{4}$$

where $f(\mathbf{X})$ is the objective function, M is the total number of particles.

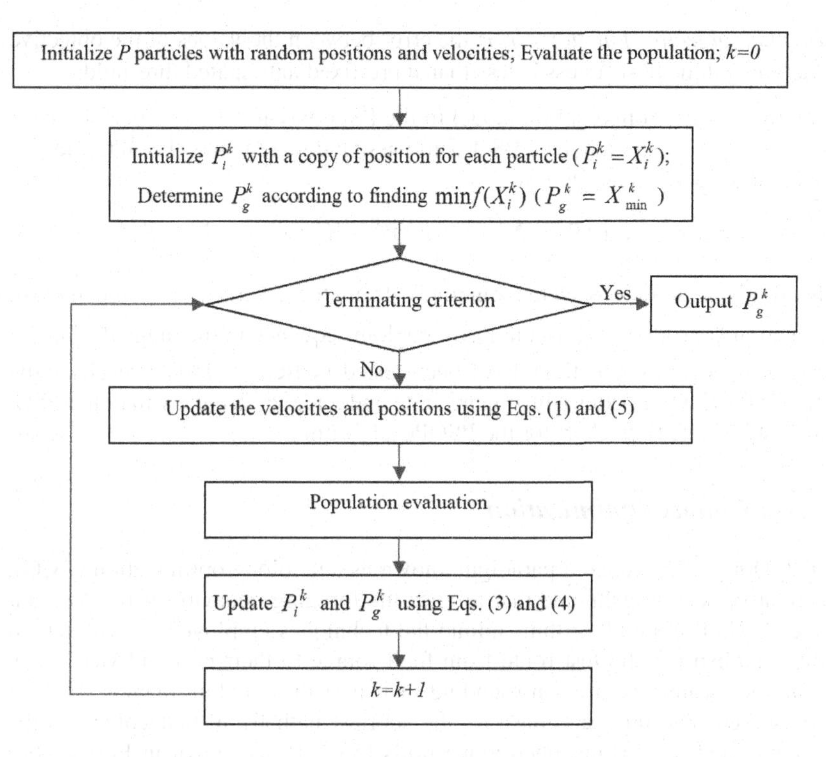

Fig. 1 The flow chart for the PSOPC algorithm

The pseudo-code of the PSO algorithm can be summarized as follows:

Step 1: *Initialization.* Initialize an array of particles with random positions and their associated velocities.

Step 2: *Function evaluation.* Evaluate the fitness function of each particle.

Step 3: *Local best updating.* Compare the current value of the fitness function with the particles' previous best value and update $\mathbf{P}_i^k$ according to Eq. (3).

Step 4: *Global best updating.* Determine the current global minimum fitness value among the current positions and update $\mathbf{P}_g^k$ according to Eq. (4).

Step 5: *Solution construction.* Change the velocities according to Eq. (2) and move each particle to the new position according to Eq. (1).

Step 6: *Terminating criterion controlling.* Repeat Steps 2–5 until a terminating criterion is satisfied. The terminating criteria are usually one of the following:

- *Maximum number of iterations*: the optimization process is terminated after a fixed number of iterations, for example, 1000 iterations.
- *Number of iterations without improvement*: the optimization process is terminated after some fixed number of iterations without any improvement.

- *Minimum objective function error*: the error between the values of the objective function and the best fitness is less than a pre-fixed anticipated threshold.

Adding the passive congregation model to the PSO may increase its performance. He *et al.* [52] proposed a hybrid PSO with passive congregation (PSOPC). In this method, the velocity is defined as

$$\mathbf{V}_i^{k+1} = \omega \mathbf{V}_i^k + c_1 r_1 (\mathbf{P}_i^k - \mathbf{X}_i^k) + c_2 r_2 (\mathbf{P}_g^k - \mathbf{X}_i^k) + c_3 r_3 (\mathbf{R}_i^k - \mathbf{X}_i^k) \tag{5}$$

where $\mathbf{R}_i$ is a particle selected randomly from the swarm; c_3 is the passive congregation coefficient; r_3 is a uniform random sequence in the range (0, 1).

Several benchmark functions have been tested in Ref. [52]. The results show that the PSOPC has a better convergence rate and a higher accuracy than the PSO. Figure 1 shows the flow chart for the PSOPC algorithm.

2.2 Ant Colony Optimization

In 1992, Dorigo developed a paradigm known as ant colony optimization (ACO), a cooperative search technique that mimics the foraging behavior of real live ant colonies [37]. The ant algorithms mimic the techniques employed by real ants to rapidly establish the shortest route from food source to their nest and vice versa. Ants start searching the area surrounding their nest in a random manner. Ethologists observed that ants can construct the shortest path from their colony to the feed source and back using pheromone trails [53, 54], as shown in Figure 2(a). When ants encounter an obstacle (Figure 2(b)), at first, there is an equal probability for all ants to move right or left, but after a while (Figure 2(c)), the number of ants choosing the shorter path increases because of the increase in the amount of the pheromone on that path. With the increase in the number of ants and pheromone on the shorter path, all of the ants will choose and move along the shorter path, Figure 2(d).

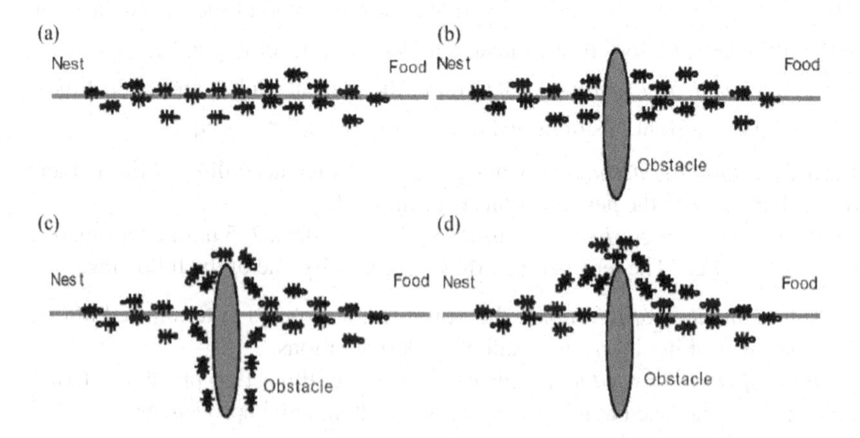

Fig. 2 Ants find the shortest path around an obstacle

In fact, real ants use their pheromone trails as a medium for communication of information among them. When an isolated ant comes across some food source in its random sojourn, it deposits a quantity of pheromone on that location. Other randomly moving ants in the neighborhood can detect this marked pheromone trail. Further, they follow this trail with a very high degree of probability and simultaneously enhance the trail by depositing their own pheromone. More and more ants follow the pheromone rich trail and the probability of the trail being followed by other ants is further enhanced by the increased trail deposition. This is an autocatalytic (positive feedback) process which favors the path along which more ants previously traversed. The ant algorithms are based on the indirect communication capabilities of the ants. In ACO algorithms, virtual ants are deputed to generate rules by using heuristic information or visibility and the principle of indirect pheromone communication capabilities for iterative improvement of rules.

ACO was initially used to solve the traveling salesman problem (TSP). The aim of TSP is finding the shortest Hamiltonian graph, $G=(N,E)$, where N denotes the set of nodes, and E is the set of edges. The general procedure of the ACO algorithm manages the scheduling of four steps [3]:

Step 1: *Initialization.* The initialization of the ACO includes two parts: the first consists mainly of the initialization of the pheromone trail. Second, a number of ants are arbitrarily placed on the nodes chosen randomly. Then each of the distributed ants will perform a tour on the graph by constructing a path according to the node transition rule described below.

Step 2: *Solution construction.* Each ant constructs a complete solution to the problem according to a probabilistic state transition rule. The state transition rule depends mainly on the state of the pheromone and visibility of ants. Visibility is an additional element used to make this method more efficient. For the path between i to j, it is represented as η_{ij} and in TSP, it has a reverse relation with the distance between i to j. The node transition rule is probabilistic. For the kth ant on node i, the selection of the next node j to follow is according to the node transition probability

$$P_{ij}(t) = \frac{[\tau_{il}(t)]^{\alpha} \cdot [\eta_{il}]^{\beta}}{\sum_{l \in N_i^k} [\tau_{il}(t)]^{\alpha} \cdot [\eta_{il}]^{\beta}} \quad \forall j \in N_i^k \tag{6}$$

where $\tau_{ij}(t)$ is the intensity of pheromone laid on edge (i, j); N_i^k is the list of neighboring nodes from node i available to ant k at time t. Parameters α and β represent constants which control the relative contribution between the intensity of pheromone laid on edge (i, j) reflecting the previous experiences of the ants about this edge, and the value of visibility determined by a Greedy heuristic for the original problem.

Step 3: *Pheromone updating rule.* When every ant has constructed a solution, the intensity of pheromone trails on each edge is updated by the pheromone updating rule. The pheromone updating rule is applied in two phases. First, an evaporation phase where a fraction of the pheromone evaporates, and then a reinforcement phase when the elitist ant which has the best solution among others, deposits an amount of pheromone

$$\tau_{ij}(t+n) = (1-\rho)\cdot\tau_{ij}(t) + \rho\cdot\Delta\tau_{ij}^{+} \tag{7}$$

where ρ $(0<\rho<1)$ represents the persistence of pheromone trails ($(1-\rho)$ is the evaporation rate); n is the number of variables or movements an ant must take to complete a tour and $\Delta\tau_{ij}^{+}$ is the amount of pheromone increase for the elitist ant and equals

$$\Delta\tau_{ij}^{+} = \frac{1}{L^{+}} \tag{8}$$

where L^{+} is the length of the solution found by the elitist ant.

Step 4: *Terminating criterion controlling.* Steps 2 and 3 are iterated until a terminating criterion.

The flow chart of the ACO procedure is illustrated in Figure 3.

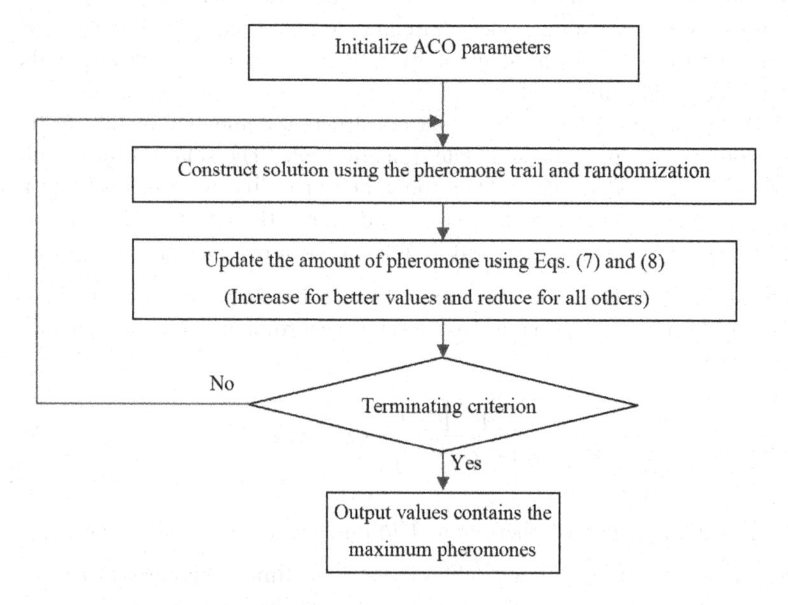

Fig. 3 The flow chart for the ACO algorithm

2.3 Harmony Search Algorithm

Harmony search (HS) algorithm is based on musical performance processes that occur when a musician searches for a better state of harmony, such as during jazz improvisation [45]. The engineers seek for a global solution as determined by an objective function, just like the musicians seek to find musically pleasing harmony as determined by an aesthetic. The HS algorithm was presented in previous chapters, and we briefly explain the steps in the algorithm here. Figure 4 shows the HS optimization procedure including the following steps [1]:

Step 1: *Initialization.* HS algorithm includes a number of optimization operators, such as the harmony memory (**HM**), the harmony memory size (HMS), the harmony memory considering rate (HMCR), and the pitch adjusting rate (PAR). In the HS algorithm, the **HM** stores the feasible vectors, which are all in the feasible space. The harmony memory size determines the number of vectors to be stored.

$$\mathbf{HM} = \begin{bmatrix} \mathbf{X}^1 \\ \vdots \\ \mathbf{X}^{HMS} \end{bmatrix}_{HMS \times ng} \tag{9}$$

Step 2: *Solution construction.* A new harmony vector is generated from the **HM**, based on memory considerations, pitch adjustments, and randomization. The HMCR varying between 0 and 1 sets the rate of choosing a value in the new vector from the historic values stored in the **HM**, and (1–HMCR) sets the rate of randomly choosing one value from the possible range of values.

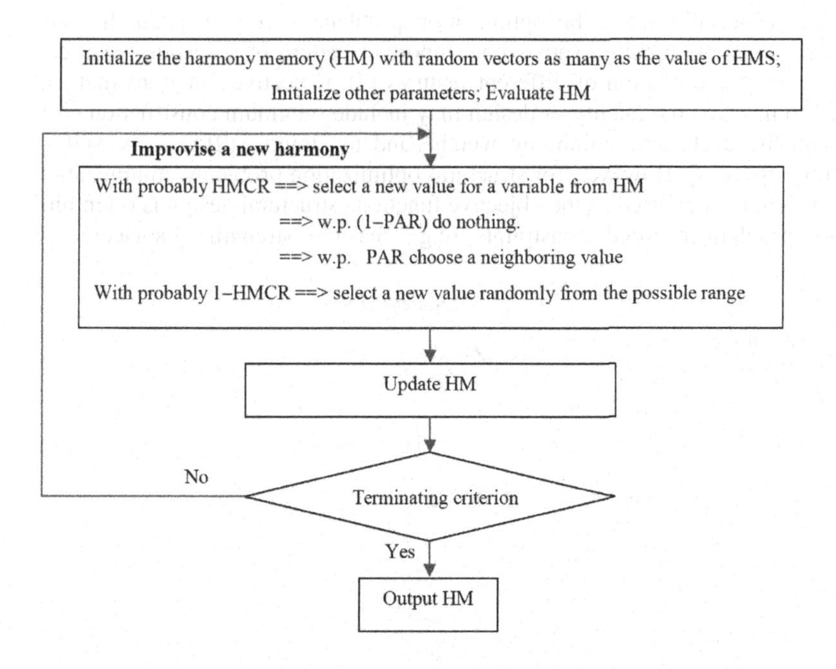

Fig. 4 The flow chart for the HS algorithm

$$x_i^k = \begin{cases} \text{Select from } \{x_i^1, x_i^2, \ldots, x_i^{HMS}\} & \text{w.p. HMCR} \\ \text{Select from the possible range} & \text{w.p. } (1-\text{HMCR}) \end{cases} \tag{10}$$

where x_i^k is the ith design variable in the iteration k, and w.p. is abbreviation for "with probability". The pitch adjusting process is performed only after a value is chosen from the **HM**. The value (1–PAR) sets the rate of doing nothing. A PAR of 0.1 indicates that the algorithm will choose a neighboring value with 10% ×HMCR probability.

Step 3: *Harmony memory updating.* In Step 3, if a new harmony vector is better than the worst harmony in the **HM**, judged in terms of the objective function value, the new harmony is included in the **HM** and the existing worst harmony is excluded from the **HM**.

Step 4: *Terminating criterion controlling.* Repeat Steps 2 and 3 until the terminating criterion is satisfied. The computations are terminated when the terminating criterion is satisfied. Otherwise, Steps 2 and 3 are repeated.

3 Statement of the Optimization Design Problem

Selection of the objective function in optimal design problems is highly significant. Usually finding a mathematical formula for the objective function is not an easy task, especially when the optimization problem is very detailed. In most cases, the objective function shows one important feature of a design, but it can also contain a combination of different features [2]. Objective functions that can be used to measure the quality of design may include minimum construction cost, minimum life cycle cost, minimum weight, and maximum stiffness, as well as other objectives [1]. However, for structural optimization problems, minimization of the weight is often used as the objective function. Structural design is often limited by problem-specified constraints (e.g., feasible strength, displacements,

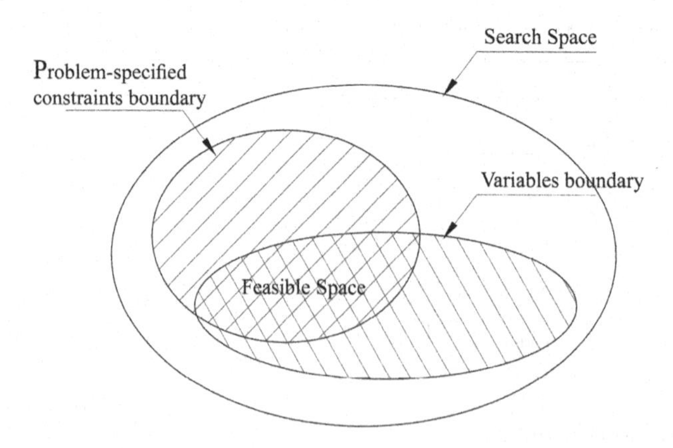

Fig. 5 Search space division

eigen-frequencies) and design variable constraints (e.g., type and size of the available structural members and cross-sections). The optimum design of structures involves a set of design variables that has the minimum weight located in the feasible space which does not violate either problem-specified constraints or design variable constraints, as illustrated in Figure 5.

3.1 Optimum Design of Truss Structures

Optimum design of truss structures involves arriving at optimum values for member cross-sectional areas x_i that minimize the structural weight W.

$$\begin{aligned} &\text{Find} \quad \mathbf{X} = [x_1, x_2, \ldots, x_{ng}], \\ &\qquad\quad x_i \in D_i \\ &\text{to minimize} \quad W(\mathbf{X}) = \sum_{i=1}^{nm} \gamma_i \cdot x_i \cdot L_i \end{aligned} \tag{11}$$

where $\mathbf{X}$ is the vector containing the design variables; ng is the number of design variables or the number of groups; $W(\mathbf{X})$ is the cost function which is taken as the weight of the structure; nm is the number of members making up the structure; γ_i is the material density of member i; L_i is the length of member i; D_i is an allowable set of values for the design variable x_i which can be considered as a continuous set or a discrete one. In the continuous problems, the design variables can vary continuously in the optimization

$$D_i = \{x_i \mid x_i \in [x_{i,\min}, x_{i,\max}]\} \tag{12}$$

where $x_{i,\min}$ and $x_{i,\max}$ are minimum and maximum allowable values for the design variable i, respectively. If the design variables represent a selection from a set of parts, the problem is considered as discrete

$$D_i = \{d_{i,1}, d_{i,2}, \ldots, d_{i,r(i)}\} \tag{13}$$

where $r(i)$ is the number of available discrete values for the ith design variable.

This minimum design also has to satisfy the problem-specified constraints that limit structural responses, as follows

$$\begin{aligned} &\text{subject to} \quad \delta_{\min} \le \delta_i \le \delta_{\max} \qquad i = 1,2,\ldots,m \\ &\qquad\qquad\quad \sigma_{\min} \le \sigma_i \le \sigma_{\max} \qquad i = 1,2,\ldots,nm \end{aligned} \tag{14}$$

where σ_i and δ_i are the stress and nodal deflection, respectively; m is the number of nodes; and min and max mean the lower and upper boundaries, respectively.

3.2 Optimum Design of Steel Frames

Similar to truss structures, the aim of the optimum design of steel frames is to find a design with minimum weight as described in Equation (11) which must satisfy the following constraints:

Stress constraints

$$\sigma_{\min} \le \sigma_i \le \sigma_{\max} \qquad i = 1,2,...,nm \tag{15}$$

Maximum lateral displacement

$$\frac{\Delta_T}{H} \le R \tag{16}$$

Inter-story displacement constraints

$$\frac{\Delta_j}{h_j} \le R_I \qquad j = 1,2,...,ns \tag{17}$$

where D_i is considered a set of 267 W-sections from the AISC database [55] for the design variable x_i; Δ_T is the maximum lateral displacement; H is the height of the frame structure; R is the maximum drift index; Δ_j is the inter-story drift; h_j is the story height of the jth floor; ns is the total number of stories; and R_I is the inter-story drift index permitted by the code of the practice.

For the code of practice AISC [55], the allowed inter-story drift index is given as 1/300, and the LRFD interaction formula constraints (AISC, Equation H1-1a,b) are expressed as

$$\frac{P_u}{2\phi_c P_n} + \left(\frac{M_{ux}}{\phi_b M_{nx}} + \frac{M_{uy}}{\phi_b M_{ny}} \right) \le 1 \quad \text{For} \qquad \frac{P_u}{\phi_c P_n} < 0.2 \tag{18}$$

$$\frac{P_u}{\phi_c P_n} + \frac{8}{9}\left(\frac{M_{ux}}{\phi_b M_{nx}} + \frac{M_{uy}}{\phi_b M_{ny}} \right) \le 1 \quad \text{For} \qquad \frac{P_u}{\phi_c P_n} \ge 0.2 \tag{19}$$

where P_u is the required strength (tension or compression); P_n is the nominal axial strength (tension or compression); ϕ_c is the resistance factor ($\phi_c = 0.9$ for tension, $\phi_c = 0.85$ for compression); M_{ux} and M_{uy} are the required flexural strengths in the x and y directions, respectively; M_{nx} and M_{ny} are the nominal flexural strengths in the x and y directions (for two-dimensional structures, M_{ny}=0); and ϕ_b is the flexural resistance reduction factor ($\phi_b = 0.90$).

4 A Heuristic Particle Swarm Ant Colony Optimization

The heuristic particle swarm ant colony optimization (HPSACO), a hybridized approach based on HS, PSO and ACO, is described in this section. HPSACO utilizes a particle swarm optimization with passive congregation (PSOPC) algorithm as a global search, and the ant colony approach worked as a local search. In the HPSACO algorithm, fly-back mechanism and the harmony search are used to handle the constraints. Fly-back mechanism handles the problem-specific constraints, and the HS deals with the variable constraints. HPSACO utilizes an efficient terminating criterion considering exactitude of the solutions. This terminating criterion is defined in a way that after decreasing the movements of particles, the search process stops. In the discrete method of HPSACO, agents are not allowed to select any value except discrete cross sections from the permissible list.

4.1 Combining PSO with ACO

Compared to other evolutionary algorithms based on heuristics, the advantages of PSO include an easy implementation and its smaller number of parameters to be adjusted. However, it is known that the original PSO had difficulties in controlling the balance between exploration (global investigation of the search place) and exploitation (the fine search around a local optimum) [36]. In order to improve upon this character of PSO, one method is to hybridize PSO with other approaches such as ACO. The resulted method, called particle swarm ant colony optimization (PSACO), was initially introduced by Shelokar *et al.* [56] for solving the continuous unconstrained problems and recently utilized for the design of structures by authors [57, 58]. We have applied PSOPC instead of the PSO to improve the performance of the new method. The relation of the standard deviation in ACO stage is different with Ref. [56] and the inertia weight is changed in PSOPC stage.

The implementation of PSACO algorithm consists of two stages [57]. In the first stage, it applies PSOPC, while ACO is implemented in the second stage. ACO works as a local search, wherein, ants apply pheromone-guided mechanism to refine the positions found by particles in the PSOPC stage. In the PSACO, a simple pheromone-guided mechanism of the ACO is proposed to be applied for the local search. The proposed ACO algorithm handles M ants equal to the number of particles in PSOPC.

In ACO stage, each ant generates a solution around $\mathbf{P}_g^k$ which can be expressed as

$$\mathbf{Z}_i^k = N(\mathbf{P}_g^k, \sigma) \tag{20}$$

In the above equation, $N(P_g^k, \sigma)$ denotes a random number normally distributed with mean value P_g^k and variance σ , where

$$\sigma = (x_{\max} - x_{\min}) \times \eta \tag{21}$$

η is used to control the step size. The normal distribution with mean $\mathbf{P}_g^k$ can be considered as a continuous pheromone which has the maximum value in $\mathbf{P}_g^k$ and which decreases going away from it. In ACO algorithms, the probability of selecting a path with more pheromone is greater than other paths. Similarly, in the normal distribution, the probability of selecting a solution in the neighborhood of $\mathbf{P}_g^k$ is greater than the others. This principle is used in the PSACO algorithm as a helping factor to guide the exploration and to increase the controlling in exploitation.

In the present method, the objective function value $f(\mathbf{Z}_i^k)$ is computed and the current position of ant *i*, $\mathbf{Z}_i^k$, is replaced by the current position of particle *i* in the swarm, $\mathbf{X}_i^k$, if $f(\mathbf{X}_i^k) > f(\mathbf{Z}_i^k)$ and the current ant is in the feasible space.

4.2 HS Added to PSACO as a Variable Constraint Handling Approach

The heuristic particle swarm ant colony optimization algorithm (HPSACO) is resulted from combining PSACO and HS [59]. The framework of the HPSACO algorithm is illustrated in Figure 6. A hybrid particle swarm optimizer and harmony search scheme (HPSO) was proposed by Li *et al.* [32] for truss design. A particle in the search space may violate either the problem-specific constraints or the limits of the variables as illustrated in Figure 5. If a particle flies out of the variable boundaries, the solution cannot be used even if the problem-specific constraints are satisfied. Using the harmony search-based handling approach, this problem is dealt with. In this mechanism, any component of the solution vector (particle) violating the variable boundaries can be generated randomly from $\mathbf{P}_i^k$ as

$$x_{i,j} = \begin{cases} \text{w.p. HMCR} \Longrightarrow \text{select a new value for a variable from } \mathbf{P}_i^k \\ \qquad \Longrightarrow \text{w.p. } (1-\text{PAR}) \text{ do nothing} \\ \qquad \Longrightarrow \text{w.p. PAR choose a neighboring value} \\ \text{w.p. } (1-\text{HMCR}) \Longrightarrow \text{select a new value randomly} \end{cases} \tag{22}$$

where $x_{i,j}$ is the *j*th component of the particle i The HMCR varying between 0 and 1 sets the rate of choosing a value in the new vector from the historic values stored in the $\mathbf{P}_i^k$, and (1–HMCR) sets the rate of randomly choosing one value from the possible list of values. The pitch adjusting process is performed only after a value is chosen from $\mathbf{P}_i^k$. The value (1–PAR) sets the rate of doing nothing. A PAR (Pitch Adjusting Rate) of 0.1 indicates that the algorithm will choose a neighboring value with 10% ×HMCR probability. Therefore, the harmony search concept is used to check whether the particles violate the variables' boundaries.

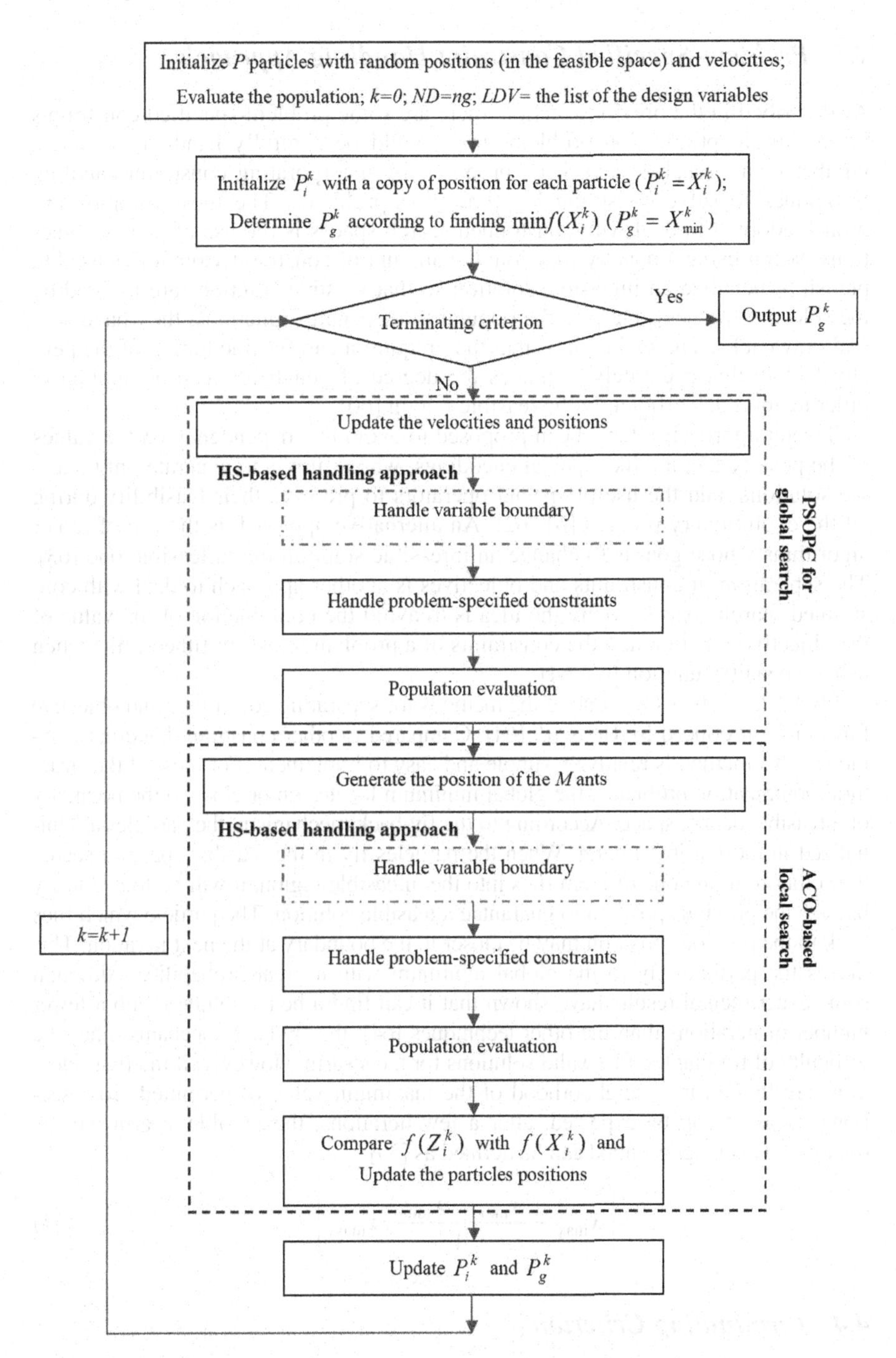

Fig. 6 The flow chart for the HPSACO

4.3 Problem-Specified Constraint Handling Approach

As described in the previous section, there are some problem-specified constraints in structural optimization problems that should be carefully handled. So far, a number of approaches have been proposed by incorporating constraint-handling techniques to solve constrained optimization problems. The most common approach adopted to deal with constrained search spaces is the use of penalty functions. When using a penalty function, the amount of constraint violation is used to punish or penalize an infeasible solution so that feasible solutions are favored by the selection process. Despite the popularity of penalty functions, they have several drawbacks. The main one is that they require a careful fine tuning of the penalty factors that accurately estimates the degree of penalization to be applied in order to approach efficiently the feasible region [60].

Several approaches have been proposed to avoid this dependency on the values of the penalty factors, like special encodings, whose aim is to generate only feasible solutions, and the use of special operators to preserve their feasibility during all the evolutionary process [61, 62]. An alternative approach is the use of repair algorithms whose goal is to change an infeasible solution into a feasible one [63]. The separation of constraints and objectives is another approach to deal with constrained search spaces, where the idea is to avoid the combination of the value of the objective function and the constraints of a problem to assign fitness, like when using a penalty function [60, 64].

Fly-back mechanism is one of the methods for separating constraints and objective functions, introduced by He *et al.* [64]. Compared to other constraint-handling techniques, this method is relatively simple and easy to implement. For most of the structural optimization problems, the global minimum locates on or close to the boundary of a feasible design space. According to the fly-back mechanism, the particles are initialized in the feasible region. When the particles fly in the feasible space to search the solution, if any one of them flies into the infeasible region, it will be forced to fly back to the previous position to guarantee a feasible solution. The particle which flies back to the previous position may be closer to the boundary at the next iteration. This makes the particles fly to the global minimum with a great probability. Although some experimental results have shown that it can find a better solution with a fewer number of iterations than the other techniques [64], the fly-back mechanism has the difficulty of finding the first valid solutions for the swarm. However, if the first selections are limited to a neighborhood of the maximum value of permitted cross sectional areas, it can be expected, after a few iterations, the feasible swarm will be obtained. This neighborhood can be defined as [59]

$$\left[x_{\max} - \frac{x_{\max} - x_{\min}}{4}, x_{\max} \right] \tag{23}$$

4.4 Terminating Criterion

The maximum number of the iterations is the most usual terminating criterion in PSO literature. If it is selected as a big number, the number of analyses and as a

Table 1 The pseudo-code for the HPSACO

```
Set k=0; ND= the number of design variables (ng);
Set LDV= the list of design variables;
Randomly initialize positions and velocities of all particles (from the range of [A_min, A_max])
FOR(each particle i in the initial population)
   WHILE(the constraints are violated)
      Randomly re-generate the current particle X_i
   END WHILE
   Generate local best: Set P_i^k = X_i^k
   Generate global best: Find min f(X_i^k), P_g^k is set to the position of X_min^k
END FOR

WHILE( ND≠0)
   FOR(each particle (ant) i in the swarm(colony))
      Generate the velocity and update the position of the current particle (vector) X_i^k
      Variable boundary handling: Check whether each component of the current vector violates its
            corresponding boundary or not. If it does, select the corresponding component of the
            vector from P_i^k based on memory considerations, pitch adjustments, and randomizatior
      Problem-specified constraints handling: Check whether the current particle violates the problem
            constraints or not. If it does, reset it to the previous position X_i^(k-1)
      Calculate the fitness value f(X_i^k) of the current particle
      Generate the position of the current ant Z_i^k = N(P_g^k, σ)
      Variable boundary handling: Check whether each component of the current vector violates its
            corresponding boundary or not. If it does, select the corresponding component of the
         vector from P_i^k based on memory considerations, pitch adjustments, and randomization.
      Problem-specified constraints handling: Check whether the current ant violates the problem
            constraints or not. If it does, reset it to the current particle X_i^k
      Calculate the fitness value f(Z_i^k) of the current ant
      Update current particle position: Compare the fitness value of current ant with
            current particle. If the f(Z_i^k) is better than the fitness value of f(X_i^k),
            set f(X_i^k) = f(Z_i^k) and X_i^k = Z_i^k
      Update local best: Compare the fitness value of f(P_i^k) with f(X_i^k).
            If the f(X_i^k) is better than the fitness value of f(P_i^k),
            set P_i^k to the current position X_i^k
   END FOR
   Update global best: Find the global best position in the swarm. If the f(X_i^k) is
         better than the fitness value of f(P_g^k), P_g^k is set to the position
         of the current particle X_i^k

   FOR(each j ∈ LDV)
      IF( |V_ij^k| < A*/2,   ∀i ∈ swarm )
            Set ND=ND-1 and Delete variable j from LDV
      END IF
   END FOR

   Set k =k+1
END WHILE
```

result, the time of optimization will increase; vice versa, if it is selected small, the probability of finding a desirable solution will decrease. Thus, the necessity for an exact definition of the terminating criterion is vital. The following terminating criterion is considered to fulfill this goal.

This terminating criterion is defined by using a pre-fixed value denoted by A^*. For the discrete problems, A^* is equal to the minimum value of the difference between cross-sectional areas of two successive discrete sections, and for continuous problems, A^* is considered as the required exactitude of the solutions with a reverse relation. According to this criterion, as A^* increases, exactitude of the solutions decreases and the searching process must be stopped earlier, and if the amount of A^* decreases, then the searching process must be continued until an exact result is attained. Therefore, if in an iteration of search process, the absolute value of the component i in all of the particles' velocity vectors is less than $A^*/2$, continuation of the search process cannot change the amount of variable i; then the variable i reaches an optimum value and can be deleted from the virtual list of design variables. As a result, the terminating criterion is defined as continuing the search process until all variables are deleted. In the other words, when the variation of a variable is less than $A^*/2$, this criterion omits it from the virtual list of variables. When this list is emptied, the search process stops. With these alterations, the number of iterations decreases.

The pseudo-code for the HPSACO algorithm using this terminating criterion is listed in Table 1.

4.5 A Discrete HPSACO

In the discrete HPSACO, a new position of each agent is defined as the following:

For particles

$$\mathbf{X}_i^{k+1} = Fix\,(\mathbf{X}_i^k + \mathbf{V}_i^{k+1}) \tag{24}$$

For ants

$$\mathbf{Z}_i^k = Fix\left(N\,(\mathbf{P}_g^k, \sigma)\right) \tag{25}$$

where $Fix(\mathbf{X})$ is a function which rounds each element of $\mathbf{X}$ to the nearest permissible discrete value. Using this position updating formula, the agents will be permitted to select discrete values. Although this change is simple and efficient, it may reduce the exploration in the algorithm. Therefore, in order to increase the exploration, the velocity of particles is redefined [58] as

$$\begin{aligned}\mathbf{V}_i^{k+1} = \omega\mathbf{V}_i^k + c_1 r_1(\mathbf{P}_i^k - \mathbf{X}_i^k) + c_2 r_2(\mathbf{P}_g^k - \mathbf{X}_i^k) + \\ c_3 r_3(\mathbf{R}_i^k - \mathbf{X}_i^k) + c_4 r_4(\mathbf{Rd}_i^k - \mathbf{X}_i^k)\end{aligned} \tag{26}$$

where c_4 is the exploration coefficient; r_4 is a uniformly distributed random number in the range of (0, 1); and $\mathbf{Rd}^k$ is a vector generated randomly from the search domain.

4.6 Parameter Setting

For the proposed algorithm, a population of 50 individuals is used for both particles and ants (M=50); the value of constants c_1 and c_2 are set 0.8 and the passive congregation coefficient c_3 is taken as 0.6. The value of inertia weight ($\omega(k)$) decreases linearly from 0.9 to 0.4 as follows

$$\omega(k) = 0.9 - 0.0015 \times k \geq 0.4 \tag{27}$$

where k = the iteration number. In this way, the balance of $\omega(k)$ with fast rate of convergence in the HPSACO method is maintained.

The amount of step size (η) in ACO stage is recommended as 0.01 [57]. If η is too small, the velocity of particles will decrease rapidly and the search process will stop in early iterations; thus the obtained results stay far away from an optimum; on the contrary, if it is selected too big, the HPSACO algorithm will perform similar to the PSOPC algorithm and the effect of the ACO stage will be eliminated, and a desirable solution cannot be obtained in smaller number of iterations.

The parameters of the HS part (HMCR and PAR) similar to the effect of η, can be investigated. With small values for HMCR (large values for PAR), the effect of the HS part will be deleted. We have selected these values close to the amounts employed in the original HS algorithm [1]. If HMCR is selected from the range of [0.8, 0.98] and PAR is taken from [0.05,0.25], we expect a good performance for the HPSACO. In this study, HMCR is set to 0.95 and PAR is taken as 0.10.

5 Discussion on the Efficiency of the HPSACO

In order to verify the effectiveness of the HPSACO algorithm, a benchmark problem (10-bar truss) chosen from the literature is employed. In the next section, four design examples consisting of a 120-bar dome shaped truss with continuous design variables, a 582-bar space truss tower with 32 discrete design variables and a 3-bay 15-story steel frame structure are used to evaluate the numerical performance of the HPSACO algorithm in optimum design of different types of structures.

5.1 Benchmark Problem

The 10-bar truss design has become a common problem in the field of structural design for testing and verifying the efficiency of many different optimization methods. Figure 7 shows the geometry and support conditions for this 2-dimensional, cantilevered truss with the corresponding loading condition. The material density is 0.1 lb/in^3 (2767.990 kg/m^3) and the modulus of elasticity is 10,000 ksi (68,950 MPa). The members are subjected to the stress limits of ±25 ksi (172.375 MPa) and all nodes in both vertical and horizontal directions are subjected to the displacement limits of ±2.0 in (5.08 cm). There are 10 design variables in this example and a set of pseudo variables ranging from 0.1 to 35.0 in^2 (from 0.6452 cm^2 to 225.806 cm^2). A^* is considered as 0.001 for this example.

The PSO and PSOPC algorithms achieve the best solutions after 3,000 iterations (150,000 analyses) [32] and the HS algorithm reaches a solution after 20,000 analyses [1]. However, the HPSACO algorithm finds the best solution after about 426 iterations (10,650 analyses). The best weight of HPSACO is 5056.56 lb while the best results of PSO and PSOPC are 5061.00 lb, 5529.50 lb, respectively. The results of this method are compared with other methods in Table 2.

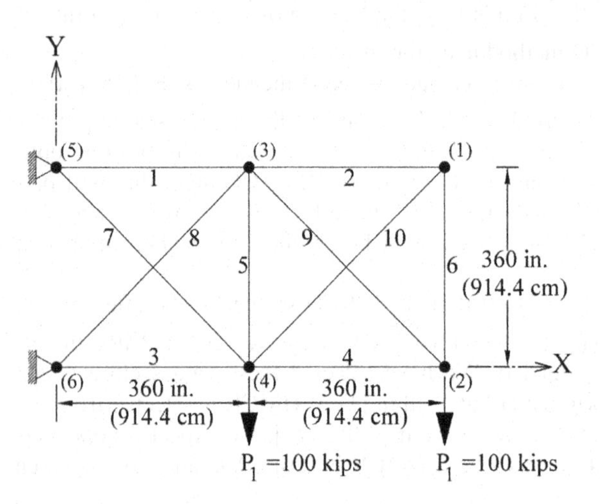

Fig. 7 A 10-bar planar truss

Table 2 Optimal design comparison for the 10-bar planner truss

Element group		Optimal cross-sectional areas (in.^{2})						
		GA [12]	HS [1]	PSO [32]	PSOPC [32]	HPSO [32]	PSACO [57]	HPSACO
1	A_1	28.92	30.15	33.469	30.569	30.704	30.068	30.307
2	A_2	0.10	0.102	0.110	0.100	0.100	0.100	0.100
3	A_3	24.07	22.71	23.177	22.974	23.167	23.207	23.434
4	A_4	13.96	15.27	15.475	15.148	15.183	15.168	15.505
5	A_5	0.10	0.102	3.649	0.100	0.100	0.100	0.100
6	A_6	0.56	0.544	0.116	0.547	0.551	0.536	0.5241
7	A_7	7.69	7.541	8.328	7.493	7.460	7.462	7.4365
8	A_8	21.95	21.56	23.340	21.159	20.978	21.228	21.079
9	A_9	22.09	21.45	23.014	21.556	21.508	21.630	21.229
10	A_{10}	0.10	0.100	0.190	0.100	0.100	0.100	0.100
Weight (lb)		5076.31	5057.88	5529.50	5061.00	5060.92	5057.36	5056.56

5.2 Discussion

The main reasons for the improvements obtained by the HPSACO method can be summarized as the following [59]:

1. *Increasing the exploitation*: In structural optimization, usually there are some local optimums in the neighborhood of a desirable solution. Thus, the probability of finding a desirable optimum increases with additional searches around the local optimums. HPSACO does extra search (exploitation) around the local optimums, and therefore obtains the desirable solution with higher probability in a smaller number of iterations.

The difference between the best and the worst results of the 10-bar truss for PSOPC in 50 tests is 365.2lb (7.21%), the average weight is 5173.45lb, and the standard deviation is 81.17lb (see Table 3). With adding the ACO principles to the PSOPC (PSACO [57]), these values are reduced to 3.2lb (0.06%), 5058.23lb, and 1.46lb, respectively. In addition, although PSO is a weak approach, applying ACO principles in PSO results in a improvement of its performance. The average weight of PSO+ACO in 50 runs is 5079.19lb, and the standard deviation is 4.76lb, which are better than PSOPC. Therefore, increasing the exploitation by applied pheromone-guided mechanism for updating the positions of the particles, not only improves the results, but also reduces the standard deviation drastically.

2. *Guiding the exploration*: Heuristic methods utilize two factors: the random search factor and the information collected from the search space during the optimization process. In early iterations, the random search factor has more power than the collective information factor, but the increase in the number of iterations gradually abates the power of the random search factor and increases the power of the collective information factor. In HPSACO, ACO stage plays an auxiliary role in rapidly increasing the collective information factor; consequently, the convergence rate increases faster.
   Although minimizing the maximum value of the velocity can make fewer particles violate the variable boundaries, it may also prevent the particles from crossing the problem-specific constraints and can cause the reduction in exploration. The harmony search-based handling approach deals with this problem.

PSOPC requires 3000 iterations to reach a solution for 10-bar truss. However, the number of required iterations to reach a solution for PSOPC+ACO (PSACO) in 50 runs on average is 635.2 iterations. Also, PSO+ACO on average needs 567 iterations to reach the optimum solution, while PSO cannot reach an appropriate solution until the maximum number of iterations is achieved (3000 iterations).

In order to investigate the advantages of the HS-based handling approach, the comparison of the performance of PSOPC with HPSO (PSOPC+HS), or PSO+ ACO with PSO+ACO+HS, or PSOPC+ACO (PSACO) with PSOPC+ACO+HS (HPSACO) can be helpful. Table 3 summarizes the performances of all the above mentioned PSO-based approaches for the 10-bar truss on 50 runs for each algorithm. Although the results and standard deviations of HPSACO and PSACO do not differ much, the convergence rate of HPSACO is higher than that of PSACO. In average, HPSACO needs 420.3 iterations to reach a solution, while for PSACO this average number is 635.2.

3. *Efficient terminating criterion:* In optimization problems, the terminating criterion is a part of the search process which can be used to eliminate additional unnecessary iterations. To fulfill this goal, an efficient terminating criterion is defined as continuing the search process until the variation of a variable is less than a pre-defined value.

Figure 8 shows the average and a typical maximum absolute value of velocity for the first design variable in 50 tests for the 10-bar truss without considering the proposed terminating criterion. As shown in the figure, generally $\max(|V_{i1}^k|)$ is a decreasing function with a slight disorder. When it gets less than A^*, there is a probability (even slight) that the values of velocities in the next iterations become more than A^*. Instead, if the upper bound of the maximum absolute value of velocities is selected as $A^*/2$, there is a small probability that particle velocities in the next iterations become more than A^* and as a result, continuing the search process cannot help to improve the results.

Table 3 Investigation on the performance of various PSO-based algorithms for the 10-bar truss in 50 runs

Algorithm	Minimum iterations	Maximum iterations	Average iterations	Best weight (lb)	Worst weight (lb)	Average weight (lb)	Standard deviation (lb)
PSOPC	3000	3000	3000	5061.00	5406.26	5173.45	81.17
PSOPC+HS	3000	3000	3000	5060.92	5103.63	5078.69	13.05
PSO+ACO	373	567	439.6	5065.23	5092.71	5079.19	4.76
PSO+ACO+HS	226	414	296.3	5065.61	5078.26	5070.86	2.87
PSOPC+ACO	619	655	635.2	5057.36	5060.61	5058.23	1.46
PSOPC+ACO+HS (HPSACO)	405	436	420.3	5056.56	5061.12	5057.66	1.42

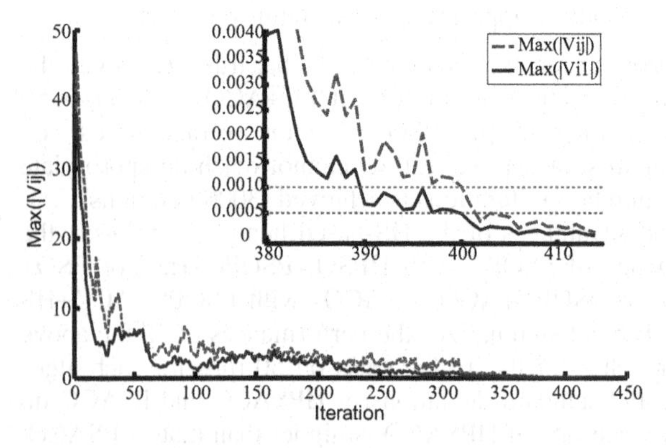

Fig. 8 The history of $\max(|V_{i1}^k|)$ in 50 tests for the 10-bar truss

6 Design Examples

6.1 A Truss Structure with Continuous Design Variables

Figure 9 shows the topology and group numbers of 120-bar dome shaped truss. The modulus of elasticity is 30,450 ksi (210,000 MPa), and the material density is 0.288 lb/in.^{3} (7971.810 kg/m^3). The yield stress of steel is taken as 58.0 ksi (400 MPa). The dome is considered to be subjected to vertical loading at all the unsupported joints. These loads are taken as −13.49 kips (−60 kN) at node 1, −6.744 kips (−30 kN) at nodes 2 through 14 and −2.248 kips (−10 kN) at the rest of the nodes. The minimum cross-sectional area of all members is 0.775 in.^{2} (2 cm^2). The allowable tensile and compressive stresses are used according to the AISC ASD [55] code, as follows

$$\begin{cases} \sigma_i^+ = 0.6F_y & \text{For } \sigma_i \geq 0 \\ \sigma_i^- & \text{For } \sigma_i < 0 \end{cases} \tag{28}$$

where σ_i^- is calculated according to the slenderness ratio:

$$\sigma_i^- = \begin{cases} \left[\left(1-\frac{\lambda_i^2}{2C_C^2}\right)Fy\right] \Big/ \left(\frac{5}{3}+\frac{3\lambda_i}{8C_C}-\frac{\lambda_i^3}{8C_C^3}\right) & \text{For } \lambda_i < C_C \\ \frac{12\pi^2 E}{23\lambda_i^2} & \text{For } \lambda_i \geq C_C \end{cases} \tag{29}$$

where E is the modulus of elasticity; F_y is the yield stress of steel; C_c is the slenderness ratio (λ_i) dividing the elastic and inelastic buckling regions ($C_C = \sqrt{2\pi^2 E/F_y}$); λ_i is the slenderness ratio ($\lambda_i = kL_i/r_i$); k is the effective length factor; L_i is the member length; and r_i is the radius of gyration which can be expressed in terms of cross-sectional areas, i.e., $r_i = aA_i^b$ [29]. Here, a and b are the constants depending on the types of sections adopted for the members. Here, pipe sections (a = 0.4993 and b = 0.6777) were used for the bars.

In this example, four cases of constraints are considered: with stress constraints and no displacement constraints (Case 1), with stress constraints and displacement limitations of ±0.1969 in (±5 mm) imposed on all nodes in x- and y-directions (Case 2), no stress constraints but displacement limitations of ±0.1969 in. (±5 mm) imposed on all nodes in z-directions (Case 3), and all constraints explained above (Case 4). For Case 1 and Case 2, the maximum cross-sectional area it is 5.0 in.^{2} (32.26 cm^2) and for Case 3 and Case 4 is 20.0 in.^{2} (129.03 cm^2).

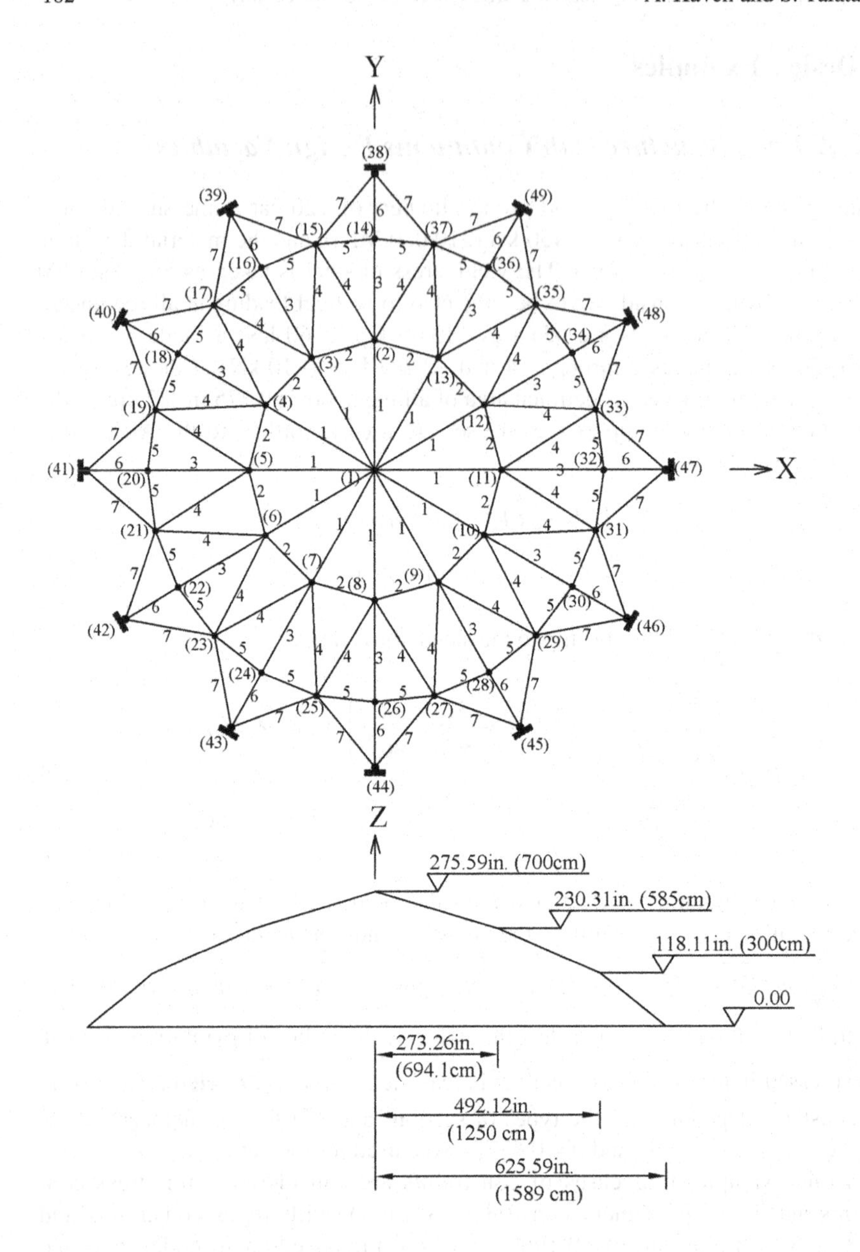

Fig. 9 A 120-bar dome shaped truss

The best solution vectors and the corresponding weights for all cases are provided in Table 4. Figure 10 shows the convergence for different cases. In all cases, HPSACO needs nearly 10,000 analyses (400 iterations) to reach a solution which is less than 125,000 (2,500 iterations) and 35,000 analyses for PSOPC and HS [1], respectively. Figures 11-14 compare the allowable and existing stress and displacement constraint values of the HPSACO results for four cases. In Case 1, the stress constraints of some elements in the 4th and 7th groups are active as shown in Figure 11(a). According to Figures 12(a), 13(a) and 14(a), the maximum values of displacements in the x, y and z directions are 0.3817in., 0.4144in. and 0.988in., respectively. In Case 2, the stress constraints in the 2nd, 4th and 7th groups and the displacement of node 26 in y direction are active. The maximum value for displacement in the x direction is 0.1817in. (Figure 11(b)).. The displacement constraints in the x and y directions do not affect the results of Case 3 and Case 4. The active constraints for Case 3 are the displacements of the 1st to 13th nodes in the z direction (Figure 14(c)). In Case 4, the stresses in the elements of the 7th group and the displacements of the 1st to 13th nodes in z directions affect the results.

Table 4 Optimal design comparison for the 120-bar dome truss (four cases)

Element group	Optimal cross-sectional areas (in.^{2})					
	Case 1			Case 2		
	PSO	PSOPC	HPSACO	PSO	PSOPC	HPSACO
1	3.147	3.235	3.311	15.978	3.083	3.779
2	6.376	3.370	3.438	9.599	3.639	3.377
3	5.957	4.116	4.147	7.467	4.095	4.125
4	4.806	2.784	2.831	2.790	2.765	2.734
5	0.775	0.777	0.775	4.324	1.776	1.609
6	13.798	3.343	3.474	3.294	3.779	3.533
7	2.452	2.454	2.551	2.479	2.438	2.539
Weight (lb)	32432.9	19618.7	19491.3	41052.7	20681.7	20078.0
	Case 3			Case 4		
	PSO	PSOPC	HPSACO	PSO	PSOPC	HPSACO
1	1.773	2.098	2.034	12.802	3.040	3.095
2	17.635	16.444	15.151	11.765	13.149	14.405
3	7.406	5.613	5.901	5.654	5.646	5.020
4	2.153	2.312	2.254	6.333	3.143	3.352
5	15.232	8.793	9.369	6.963	8.759	8.631
6	19.544	3.629	3.744	6.492	3.758	3.432
7	0.800	1.954	2.104	4.988	2.502	2.499
Weight (lb)	46893.5	31776.2	31670.0	51986.2	33481.2	33248.9

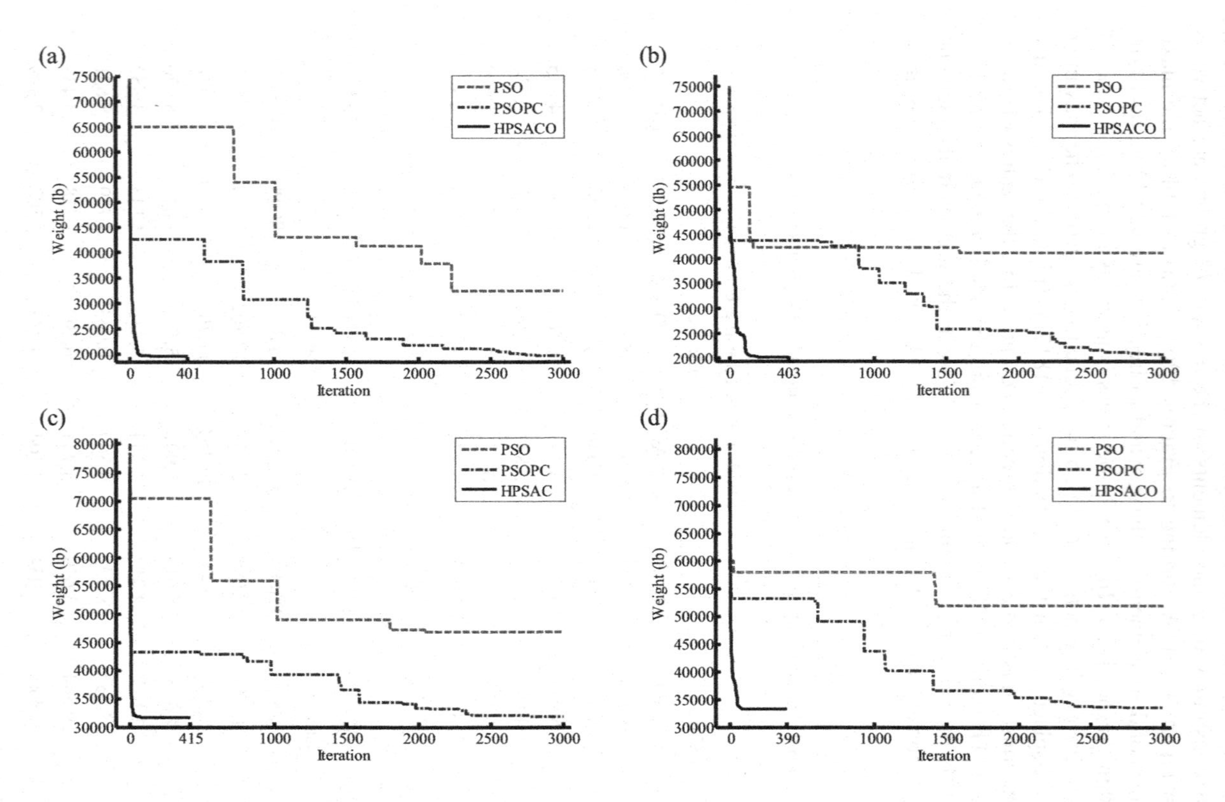

Fig. 10 Comparison of the convergence rates of the three algorithms for the 120-bar truss (a) Case 1 (b) Case 2 (c) Case 3 (d) Case 4

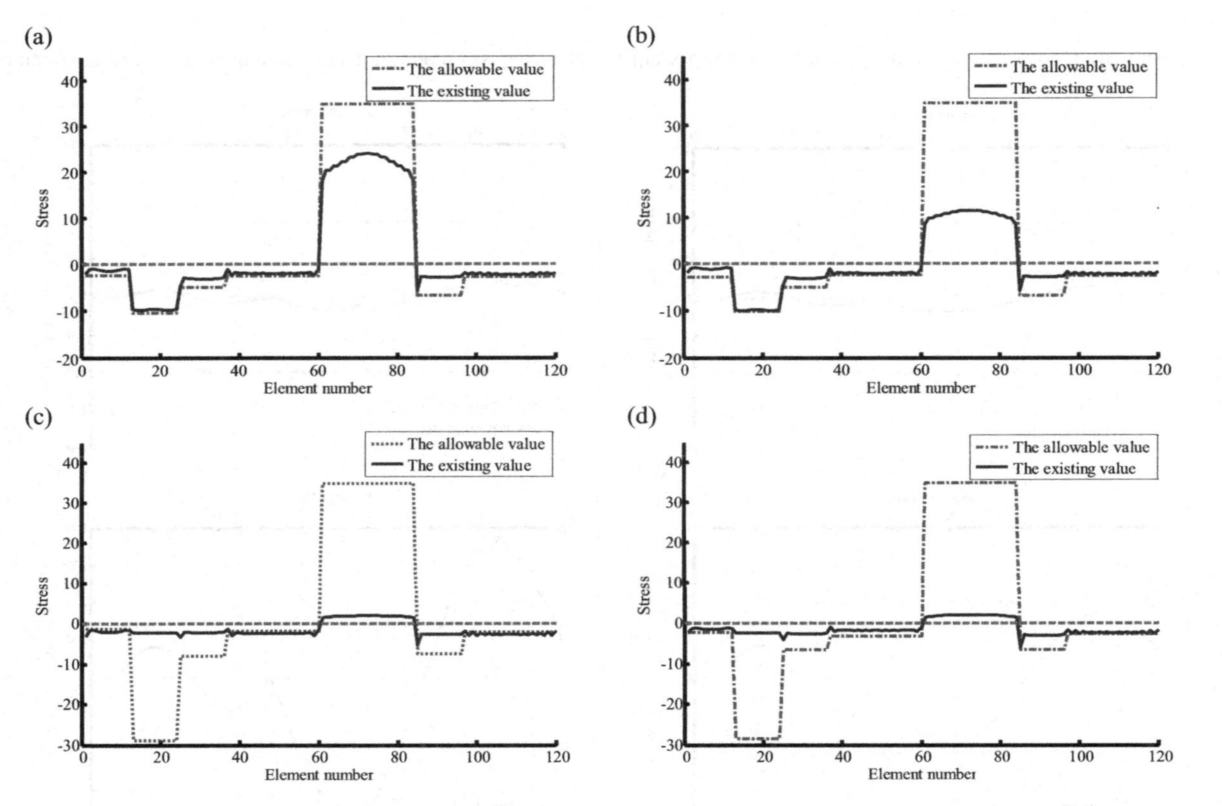

Fig. 11 Comparison of the allowable and existing stresses in the elements of the 120-bar truss (a) Case 1 (b) Case 2 (c) Case 3 (d) Case 4

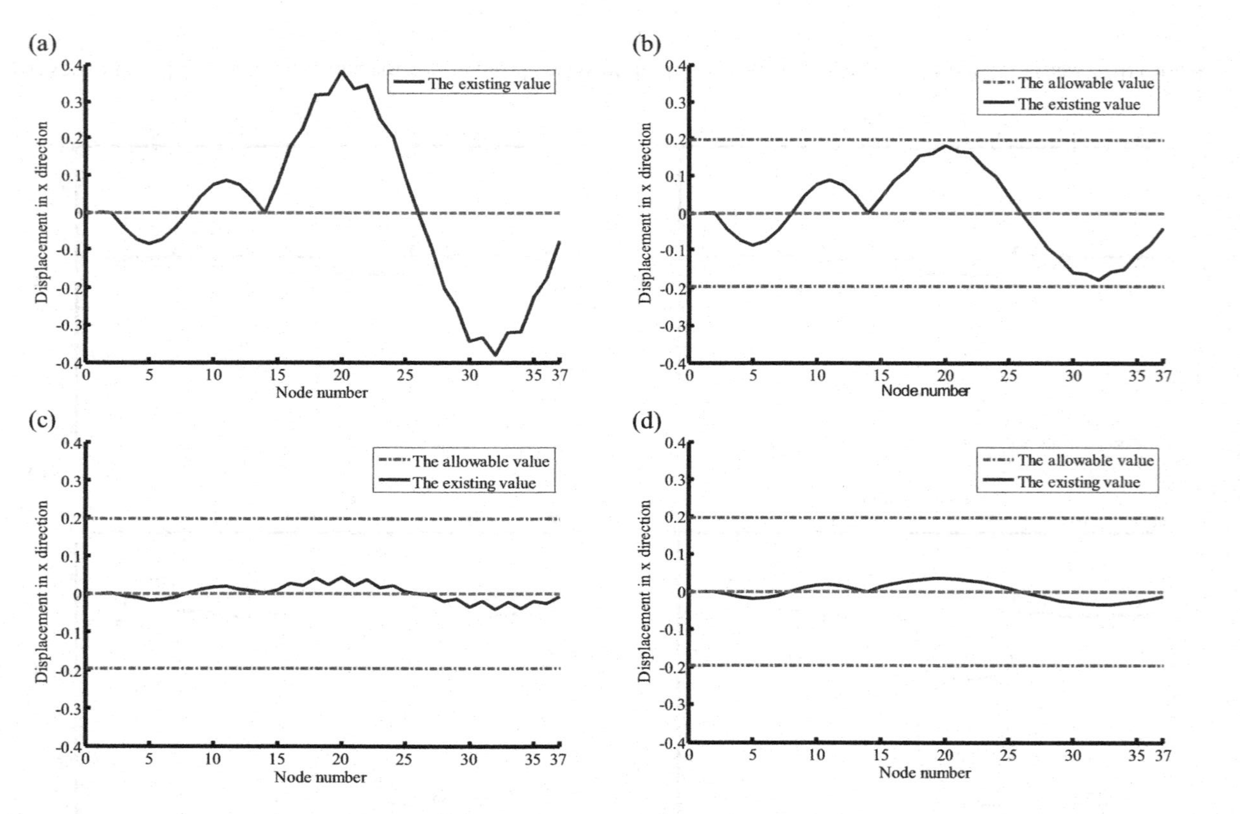

Fig. 12 Comparison of the allowable and existing nodal displacements in the x direction of the 120-bar truss (a) Case 1 (b) Case 2 (c) Case 3 (d) Case 4

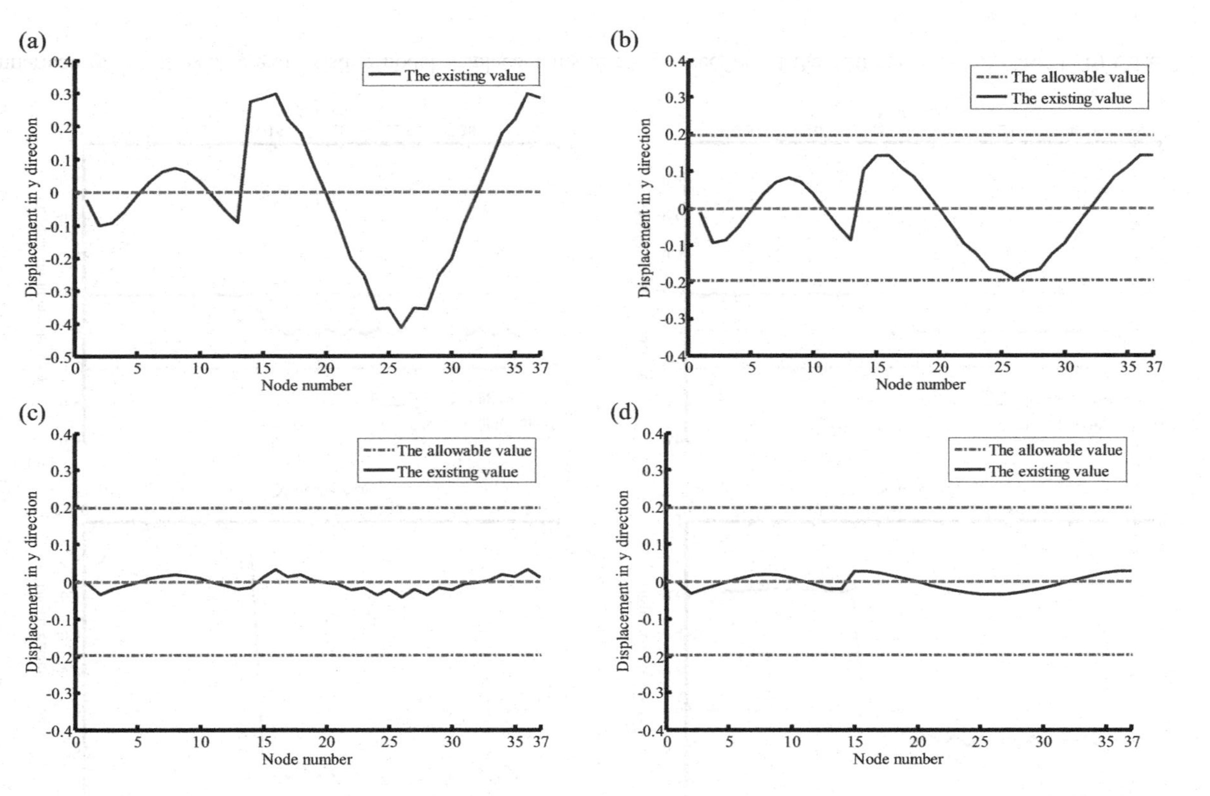

Fig. 13 Comparison of the allowable and existing nodal displacements in the y direction of the 120-bar truss (a) Case 1 (b) Case 2 (c) Case 3 (d) Case 4

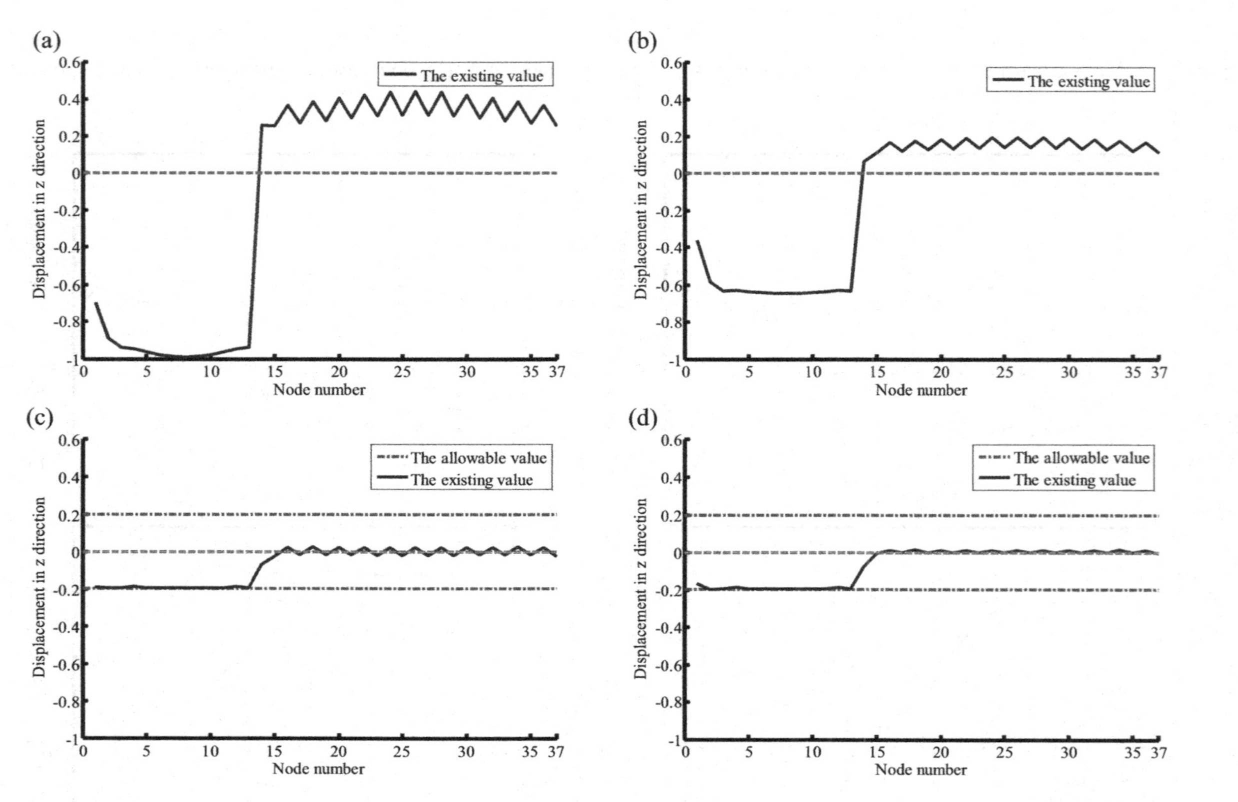

Fig. 14 Comparison of the allowable and existing nodal displacements in the z direction of the 120-bar truss (a) Case 1 (b) Case 2 (c) Case 3 (d) Case 4

6.2 A Truss Structure with Discrete Design Variables

A 582-bar tower truss shown in Figure 15 with an 80 m height is chosen from [65] as an example of truss structure with discrete design variables. The symmetry of the tower around *x*- and *y*-axes is considered to group the 582 members into 32 independent size variables. A single load case is considered consisting of lateral loads of 5.0 kN (1.12 kips) applied in both *x*- and *y*-directions and a vertical load of –30 kN (–6.74 kips) applied in the *z*-direction at all nodes of the tower. A discrete set of 137 economical standard steel sections selected from the W-shape profile list based on area and radii of gyration properties is used to size the variables [65]. The lower and upper bounds on size variables are taken as 6.16 in.^{2} (39.74 cm^2) and 215.0 in.^{2} (1387.09 cm^2), respectively. The stress limitations of the members are imposed according to the provisions of ASD-AISC, as in the previous example. The other constraint is the limitation of node displacements (no more than 8.0 cm or 3.15 in. in any direction). In addition, the maximum slenderness ratio is limited to 300 and 200 for tension and compression members, respectively.

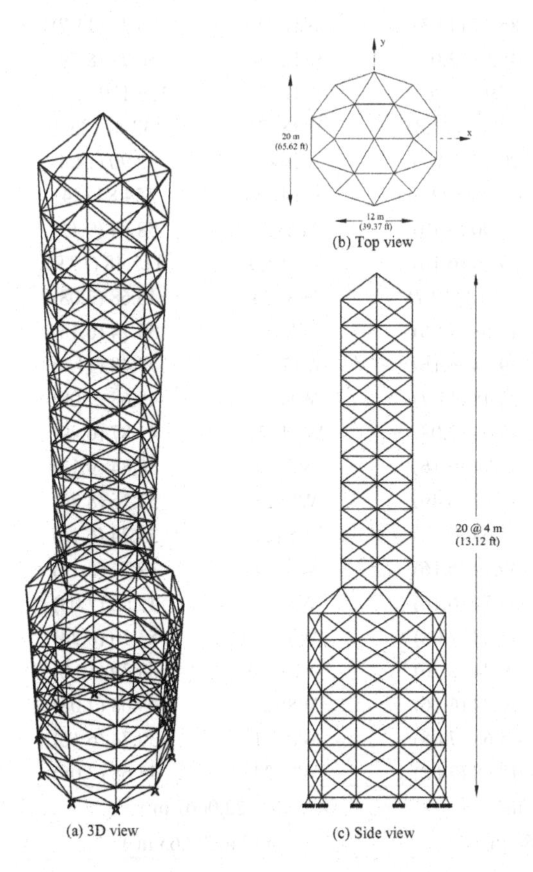

Fig. 15 A 582-bar tower truss (a) 3D view (b) Top view (c) Side view

Table 5 Optimal design comparison for the 582-bar truss tower

Element group	Optimal W-shaped sections			
	PSO [65]		HPSACO	
	Ready section	Area, cm^2 (in.^{2})	Ready section	Area, cm^2 (in.^{2})
1	W8×21	39.74 (6.16)	W8×24	45.68 (7.08)
2	W12×79	149.68 (23.2)	W12×72	136.13 (21.1)
3	W8×24	45.68 (7.08)	W8×28	53.16 (8.24)
4	W10×60	113.55 (17.08)	W12×58	109.68 (17)
5	W8×24	45.68 (7.08)	W8×24	45.68 (7.08)
6	W8×21	39.74 (6.16)	W8×24	45.68 (7.08)
7	W8×48	90.97 (14.1)	W10×49	92.90 (14.4)
8	W8×24	45.68 (7.08)	W8×24	45.68 (7.08)
9	W8×21	39.74 (6.16)	W8×24	45.68 (7.08)
10	W10×45	85.81 (13.3)	W12×40	75.48 (11.7)
11	W8×24	45.68 (7.08)	W12×30	56.71 (8.79)
12	W10×68	129.03 (20)	W12×72	136.129 (21.1)
13	W14×74	140.65 (21.8)	W18×76	143.87 (23.3)
14	W8×48	90.97 (14.1)	W10×49	92.90 (14.4)
15	W18×76	143.87 (22.3)	W14×82	154.84 (24)
16	W8×31	55.90 (9.13)	W8×31	58.84 (9.12)
17	W8×21	39.74 (6.16)	W14×61	115.48 (17.9)
18	W16×67	127.10 (19.7)	W8×24	45.68 (7.08)
19	W8×24	45.68 (7.08)	W8×21	39.74 (6.16)
20	W8×21	39.74 (6.16)	W12×40	75.48 (11.7)
21	W8×40	75.48 (11.7)	W8×24	45.68 (7.08)
22	W8×24	45.68 (7.08)	W14×22	41.87 (6.49)
23	W8×21	39.74 (6.16)	W8×31	58.84 (9.12)
24	W10×22	41.87 (6.49)	W8×28	53.16 (8.24)
25	W8×24	45.68 (7.08)	W8×21	39.74 (6.16)
26	W8×21	39.74 (6.16)	W8×21	39.74 (6.16)
27	W8×21	39.74 (6.16)	W8×24	45.68 (7.08)
28	W8×24	45.68 (7.08)	W8×28	53.16 (8.24)
29	W8×21	39.74 (6.16)	W16×36	68.39 (10.6)
30	W8×21	39.74 (6.16)	W8×24	45.68 (7.08)
31	W8×24	45.68 (7.08)	W8×21	39.74 (6.16)
32	W8×24	45.68 (7.08)	W8×24	45.68 (7.08)
Volume	22.3958 m^3 (1366674.89 in^3)		22.0607 m^3 (1346227.65 in^3)	

PSO has obtained the lightest design when compared to some other meta-heuristic algorithms such as evolution strategies, simulated annealing, tabu search, ant colony optimization, harmony search and genetic algorithms reported by Hasançebi *et al.* [65]. Table 5 gives the best solution vectors of the PSO and HPSACO algorithms [66]. The optimum result of the HPSACO approach is 22.06 m^3 while it is 22.39 m^3 for the PSO algorithm. HPSACO needs nearly 8,500 analyses to reach a solution which is significantly less than 50,000 analyses for PSO.

Figure 16 compares the allowable and existing stress ratio and displacement value in the x direction of the HPSACO result. The maximum values of displacements in the *x*, *y* and *z* directions are 3.1498 in., 2.9881 in. and 0.9258 in., respectively. The maximum stress ratio is 93.06% as show in the figure.

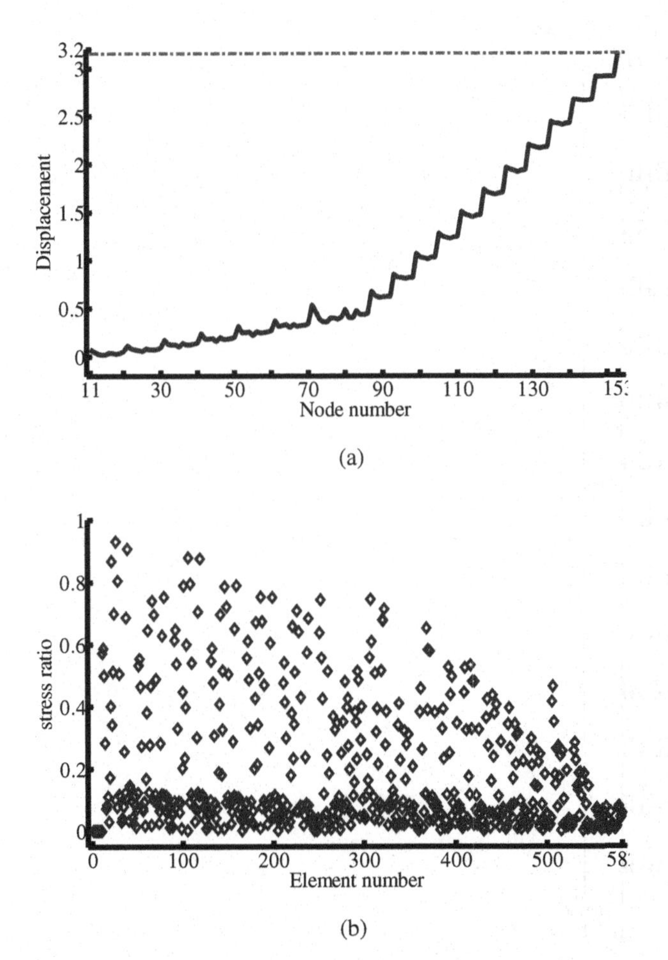

Fig. 16 Comparison of the allowable and the existing constraints for the 582-bar truss using the HPSACO (a) displacement in the x direction (b) stress ratio

6.3 A Steel Frame Structure

Figure 17 shows the configuration and applied loads of 3-bay 15-story frame structure. The displacement and AISC combined strength constraints are the performance constraint of the frame. The sway of the top story is limited to 9.25 in. (23.5 cm). The material has a modulus of elasticity E=29,000 ksi (200,000 MPa) and a yield stress of f_y=36 ksi (248.2 MPa). The effective length factors of the members are calculated as $K_x \geq 0$ for a sway-permitted frame and the out-of-plane effective length factor is specified as K_y=1.0. Each column is considered as unbraced along its length, and the unbraced length for each beam member is specified as one-fifth of the span length.

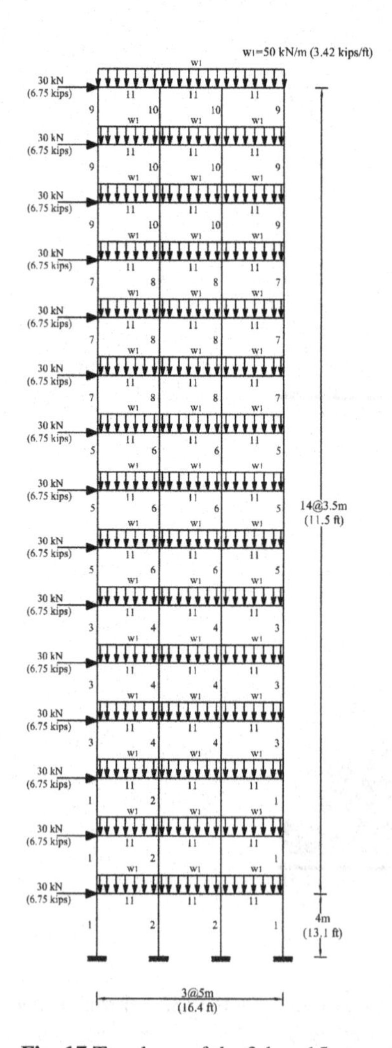

Fig. 17 Topology of the 3-bay 15-story frame

The optimum design of the frame is obtained after 6,800 analyses by using HPSACO, having the minimum weight of 426.36 kN (95.85 kips). The optimum designs for PSOPC and PSO had the weight of 452.34 kN (101.69 kips) and 496.68 kN (111.66 kips), respectively. Table 6 summarizes the optimal designs for these algorithms.

The global sway at the top story is 11.57 cm (4.56 in.), which is less than the maximum sway. Figure 18 shows the inter-story drift for each story and the stress ratio of elements for the design of the HPSACO algorithm.

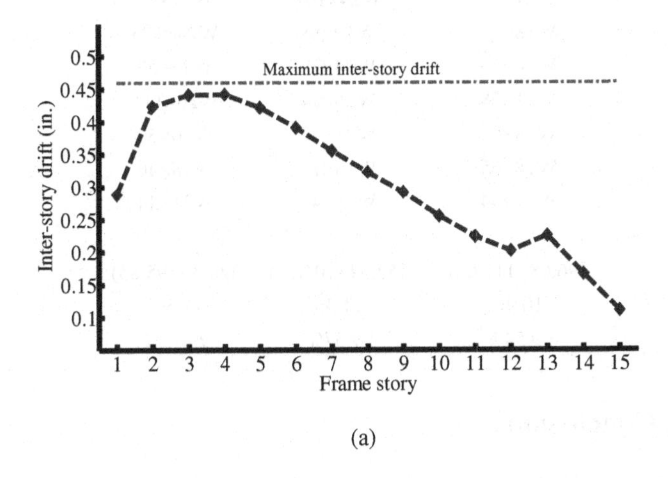

(a)

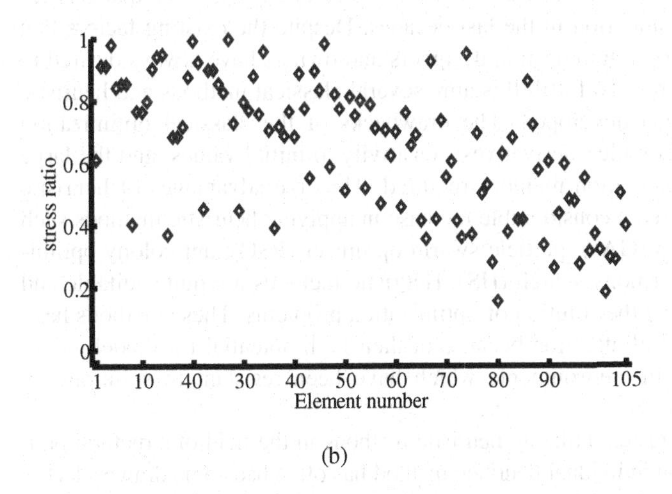

(b)

Fig. 18 Comparison of the allowable and the existing constraints for the 3-bay 15-story frame using the HPSACO (a) inter-story drift (b) stress ratio

Table 6 Optimal design comparison for the 3-bay 15-story frame

Element group	Optimal W-shaped sections		
	PSO	PSOPC	HPSACO
1	W33×118	W26×129	W21×111
2	W33×263	W24×131	W18×158
3	W24×76	W24×103	W10×88
4	W36×256	W33×141	W30×116
5	W21×73	W24×104	W21×83
6	W18×86	W10×88	W24×103
7	W18×65	W14×74	W21×55
8	W21×68	W26×94	W26×114
9	W18×60	W21×57	W10×33
10	W18×65	W18×71	W18×46
11	W21×44	W21×44	W21×44
Weight kN (kips)	496.68 (111.66)	452.34 (101.69)	426.36 (95.85)
The global sway (cm)	10.42	11.36	11.57
Max. stress ratio	99.54%	99.57%	99.72%

7 Summary and Conclusions

Structural design optimization is a critical and challenging activity that has received considerable attention in the last decades. Despite the existing factors that prevent performing optimum design, designers and owners have always desired to have optimal structures. To fulfill this aim, several classical methods and heuristic approaches have been developed. The drawbacks of the classical optimization methods consist of complex derivatives, sensitivity to initial values, and the large amount of their enumeration memory required. Thus the advantages of heuristic algorithms have caused a considerable increase in applying heuristic methods such as genetic algorithms (GAs), particle swarm optimizer (PSO), ant colony optimization (ACO) and harmony search (HS). Heuristic methods are quite suitable and powerful for obtaining the solution of optimization problems. These methods have attracted a great deal of attention because of their high potential for modeling engineering problems in environments which have been resistant to solutions by classic techniques.

There are several papers utilizing heuristic methods in the field of structural optimization, but using an individual heuristic method has often had some drawbacks because usually each method is suitable for solving only a specific group of problems and preference for a special method will differ depending on the kind of the problem being studied. One technique overcome to these problems is to hybridize various methods to reach a single robust approach. In this chapter, a new hybridized approach based on HS, PSO and ACO is presented which is called the heuristic particle swarm ant colony optimization (HPSACO).

HPSACO utilizes a particle swarm optimization with a passive congregation (PSOPC) algorithm as a global search, and the idea of an ant colony approach worked as a local search which updates the positions of the particles by applied pheromone-guided mechanism. This principle is used in the HPSACO as a helping factor to guide the exploration and to increase the control of exploitation. In the HPSACO algorithm, fly-back mechanism and the harmony search are used to handle the constraints. Fly-back mechanism handles the problem-specific constraints, and the HS deals with the variable constraints. A particle in the search space may violate either the problem-specific constraints or the limits of the variables. Since the fly-back mechanism is used to handle the problem-specific constraints, the particle will be forced to fly back to its previous position regardless whether it violates the problem-specific constraints. If it flies out of the variable boundaries, the solution cannot be used even if the problem-specific constraints are satisfied. Although minimizing the maximum value of the velocity can make fewer particles violate the variable boundaries, it may also prevent the particles from crossing the problem-specific constraints and can cause the reduction in exploration. Using a harmony search based handling approach, this problem is dealt with. According to this approach, any component of the solution vector violating the variable boundaries can be regenerated from harmony memory.

In optimization problems, and particularly in structural optimization, the number of iterations is highly important. The terminating criterion is a part of the search process which can be used to eliminate additional unnecessary iterations. HPSACO utilizes an efficient terminating criterion considering exactitude of the solutions. This terminating criterion is defined in a way that after decreasing the movements of particles, the search process stops. When the variation of a variable is less than a determined exactitude, this criterion deletes it from the virtual list of variables. When this list becomes empty, the search process stops. Using this terminating criterion, the number of required iterations decreases.

Some changes are made in order to reach a discrete version of HPSACO. In the discrete method, agents are allowed to select discrete values from the permissible list of cross sections, and if any one of agents selects another value for a design variable, the discrete HPSACO changes the amount of it with the value of the nearest discrete cross section. Although this change is simple and efficient, its effect may be to reduce exploration of the algorithm. Therefore, the formula of particles' velocity is improved by adding an exploration term.

In order to find an optimum design for different types of structures, the implementation of the HPSACO methodology is investigated. We start with truss structures considering a continuous domain as the search space. The second problem contains a large-scale truss structure with a discrete search space. Then, the efficiency of the HPSACO algorithm is investigated to find optimum design of frame structures. The results confirm that the HPSACO algorithm is quite effective in finding the optimum design of structures and can be successfully applied to structural optimization problems with continuous or discrete variables.

References

1. Lee, K.S., Geem, Z.W.: A new structural optimization method based on the harmony search algorithm. Computers and Structures 82, 781–798 (2004)
2. Kaveh, A., Farahmand Azar, B., Talatahari, S.: Ant colony optimization for design of space trusses. International Journal of Space Structures 23(3), 167–181 (2008)
3. Kaveh, A., Talatahari, S.: An improved ant colony optimization for constrained engineering design problems. Engineering Computations (accepted for publication, 2009)
4. Goldberg, D.E., Samtani, M.P.: Engineering optimization via genetic algorithm, in Will KM (Eds). In: Proceeding of the 9th Conference on Electronic Computation, ASCE, February 1986, pp. 471–482 (1986)
5. Hajela, P.: Genetic search - an approach to the non-convex optimization problem. AIAA Journal 28(71), 1205–1210 (1990)
6. Rajeev, S., Krishnamoorthy, C.S.: Discrete optimization of structures using genetic algorithms. Journal of Structural Engineering, ASCE 118(5), 1233–1250 (1992)
7. Rajeev, S., Krishnamoorthy, C.S.: Genetic algorithm-based methodologies for design optimization of trusses. Journal of Structural Engineering, ASCE 123(3), 350–358 (1997)
8. Koumousis, V.K., Georgious, P.G.: Genetic algorithms in discrete optimization of steel truss roofs. Journal of Computing in Civil Engineering, ASCE 8(3), 309–325 (1994)
9. Hajela, P., Lee, E.: Genetic algorithms in truss topological optimization. International Journal of Solids and Structures 32(22), 3341–3357 (1995)
10. Wu, S.-J., Chow, P.-T.: Integrated discrete and configuration optimization of trusses using genetic algorithms. Computers and Structures 55(4), 695–702 (1995)
11. Soh, C.K., Yang, J.: Fuzzy controlled genetic algorithm search for shape optimization. Journal of Computing in Civil Engineering, ASCE 10(2), 143–150 (1996)
12. Camp, C., Pezeshk, S., Cao, G.: Optimized design of two dimensional structures using a genetic algorithm. Journal of Structural Engineering, ASCE 124(5), 551–559 (1998)
13. Shrestha, S.M., Ghaboussi, J.: Evolution of optimization structural shapes using genetic algorithm. Journal of Structural Engineering, ASCE 124(11), 1331–1338 (1998)
14. Pezeshk, S., Camp, C.V., Chen, D.: Design of nonlinear framed structures using genetic optimization. Journal of Structural Engineering, ASCE 126(3), 382–388 (2000)
15. Erbatur, F., Hasancebi, O., Tutuncil, I., Kihc, H.: Optimal design of planar and structures with genetic algorithms. Computers and Structures 75, 209–224 (2000)
16. Coello, C.A., Christiansen, A.D.: Multiobjective optimization of trusses using genetic algorithms. Computers and Structures 75(6), 647–660 (2000)
17. Greiner, D., Winter, G., Emperador, J.M.: Optimising frame structures by different strategies of genetic algorithms. Finite Elements in Analysis and Design 37(5), 381–402 (2001)
18. Kameshki, E.S., Saka, M.P.: Optimum design of nonlinear steel frames with semi-rigid connections using a genetic algorithm. Computers and Structures 79, 1593–1604 (2001)
19. Kameshki, E.S., Saka, M.P.: Genetic algorithm based optimum bracing design of non-swaying tall frames. Journal of Constructional Steel Research 57, 1081–1097 (2001)
20. Kameshki, E.S., Saka, M.P.: Optimum geometry design of nonlinear braced domes using genetic algorithm. Computers and Structures 85(1-2), 71–79 (2007)
21. Saka, M.P.: Optimum design of pitched roof steel frames with haunched rafters by genetic algorithm. Computers and Structures 81(18-19), 1967–1978 (2003)
22. Saka, M.P.: Optimum topological design of geometrically nonlinear single layer latticed domes using coupled genetic algorithm. Computers and Structures 85(21-22), 1635–1646 (2007)

23. Kaveh, A., Kalatjari, V.: Genetic algorithm for discrete sizing optimal design of trusses using the force method. International Journal for Numerical Methods in Engineering 55, 55–72 (2002)
24. Kaveh, A., Kalatjari, V.: Topology optimization of trusses using genetic algorithm, force method, and graph theory. International Journal for Numerical Methods in Engineering 58(5), 771–791 (2003)
25. Kaveh, A., Abditehrani, A.: Design of frames using genetic algorithm, force method and graph theory. International Journal for Numerical Methods in Engineering 61, 2555–2565 (2004)
26. Kaveh, A., Khanlari, K.: Collapse load factor of planar frames using a modified Genetic algorithm. Communications in Numerical Methods in Engineering 20, 911–925 (2004)
27. Kaveh, A., Rahami, H.: Analysis, design and optimization of structures using force method and genetic algorithm. International Journal for Numerical Methods in Engineering 65(10), 1570–1584 (2006)
28. Kaveh, A., Shahrouzi, M.: Simultaneous topology and size optimization of structures by genetic algorithm using minimal length chromosome. Engineering Computations 23(6), 664–674 (2006)
29. Eberhart, R.C., Kennedy, J.: A new optimizer using particle swarm theory. In: Proceedings of the Sixth International Symposium on Micro Machine and Human Science, Nagoya, Japan (1995)
30. Kennedy, J., Eberhart, R.C., Shi, Y.: Swarm intelligence. Morgan Kaufman Publishers, San Francisco (2001)
31. Schutte, J.J., Groenwold, A.A.: Sizing design of truss structures using particle swarms. Structural and Multidisciplinary Optimization 25, 261–269 (2003)
32. Li, L.J., Huang, Z.B., Liu, F., Wu, Q.H.: A heuristic particle swarm optimizer for optimization of pin connected structures. Computers and Structures 85, 340–349 (2007)
33. Perez, R.E., Behdinan, K.: Particle swarm approach for structural design optimization. Computers and Structures 85, 1579–1588 (2007)
34. Kathiravan, R., Ganguli, R.: Strength design of composite beam using gradient and particle swarm optimization. Composite Structures 81(4), 471–479 (2007)
35. Suresh, S., Sujit, P.B., Rao, A.K.: Particle swarm optimization approach for multi-objective composite box-beam design. Composite Structures 81(4), 598–605 (2007)
36. Angeline, P.: Evolutionary optimization versus particle swarm optimization: philosophy and performance difference. In: Proceeding of the Evolutionary Programming Conference, San Diego, USA (1998)
37. Dorigo, M.: Optimization, learning and natural algorithms. PhD thesis, Dip. Elettronica e Informazione, Politecnico di Milano, Italy (1992)
38. Dorigo, M., Maniezzo, V., Colorni, A.: The Ant System: Optimization by a colony of cooperating agents. IEEE Transactions on Systems, Man, and Cybernetics-Part B 26(1), 1–13 (1996)
39. Camp, C.V., Bichon, J.: Design of space trusses using ant colony optimization. Journal of Structural Engineering, ASCE 130(5), 741–751 (2004)
40. Serra, M., Venini, P.: On some applications of ant colony optimization metaheuristic to plane truss optimization. Structural Multidisciplinary Optimization 32(6), 499–506 (2006)
41. Camp, C.V., Bichon Stovall, J.P.: Design of steel frames using ant colony optimization. Journal of Structural Engineering, ASCE 131, 369–379 (2005)
42. Kaveh, A., Shojaee, S.: Optimal design of skeletal structures using ant colony optimisation. International Journal for Numerical Methods in Engineering 70(5), 563–581 (2007)
43. Kaveh, A., Hassani, B., Shojaee, S., Tavakkoli, S.M.: Structural topology optimization using ant colony methodology. Engineering Structures 30(9), 2259–2265 (2008)
44. Kaveh, A., Shahrouzi, M.: Dynamic selective pressure using hybrid evolutionary and ant system strategies for structural optimization. International Journal for Numerical Methods in Engineering 73(4), 544–563 (2008)

45. Geem, Z.W., Kim, J.H., Loganathan, G.V.: A new heuristic optimization algorithm; harmony search. Simulation 76, 60–68 (2001)
46. Lee, K.S., Geem, Z.W.: A new meta-heuristic algorithm for continuous engineering optimization: harmony search theory and practice. Computer methods in applied mechanics and engineering 194, 3902–3933 (2005)
47. Lee, K.S., Choi, C.S.: Discrete-continuous configuration optimization methods for structures using the harmony search algorithm. Key Engineering Materials 324-245, 1293–1296 (2006)
48. Saka, M.P.: Optimum geometry design of geodesic domes using harmony search algorithm. Advances in Structural Engineering 10(6), 595–606 (2007)
49. Degertekin, S.O.: Optimum design of steel frames using harmony search algorithm. Structural and Multidisciplinary Optimization 36, 393–401 (2008)
50. Degertekin, S.O.: Harmony search algorithm for optimum design of steel frame structures: a comparative study with other optimization methods. Structural Engineering and Mechanics 29, 391–410 (2008)
51. Saka, M.P.: Optimum design of steel sway frames to BS5950 using harmony search algorithm. Journal of Constructional Steel Research 65(1), 36–43 (2009)
52. He, S., Wu, Q.H., Wen, J.Y., Saunders, J.R., Paton, R.C.: A particle swarm optimizer with passive congregation. Biosystem 78, 135–147 (2004)
53. Deneubourg, J.L., Goss, S.: Collective patterns and decision-making. Ethnology Ecology and Evolution 1, 295–311 (1989)
54. Goss, S., Beckers, R., Deneubourg, J.L., Aron, S., Pasteels, J.M.: How trail laying and trail following can solve foraging problems for ant colonies. In: Hughes, R.N. (ed.) Behavioural mechanisms in food selection, NATO-ASI Series, Berlin, vol. G 20 (1990)
55. American Institute of Steel Construction, AISC, Manual of steel construction–Load resistance factor design, 3rd edn. AISC, Chicago (1991)
56. Shelokar, P.S., Siarry, P., Jayaraman, V.K., Kulkarni, B.D.: Particle swarm and ant colony algorithms hybridized for improved continuous optimization. Applied Mathematics and Computation 188, 129–142 (2007)
57. Kaveh, A., Talatahari, S.: A hybrid particle swarm and ant colony optimization for design of truss structures. Asian Journal of Civil Engineering 9(4), 329–348 (2008)
58. Kaveh, A., Talatahari, S.: A discrete particle swarm ant colony optimization for design of steel frames. Asian Journal of Civil Engineering 9(6), 563–575 (2008)
59. Kaveh, A., Talatahari, S.: Particle swarm optimizer, ant colony strategy and harmony search scheme hybridized for optimization of truss structures. Computers and Structures 87(5–6), 267–283 (2009)
60. Coello, C.A.C.: Theoretical and numerical constraint-handling techniques used with evolutionary algorithms: a survey of the state of the art. Computer Methods in Applied Mechanics and Engineering 191, 1245–1287 (2002)
61. Michalewicz, Z.: Genetic algorithms + data structures = evolution programs, 3rd edn. Springer, Heidelberg (1996)
62. Koziel, S., Michalewicz, Z.: Evolutionary algorithms, homomorphous mappings, and constrained parameter optimization. Evolutionary Computation 7(1), 19–44 (1999)
63. Liepins, G.E., Vose, M.D.: Representational issues in genetic optimization. Journal of Experimental and Theoretical Computer Science 2(2), 4–30 (1990)
64. He, S., Prempain, E., Wu, Q.H.: An improved particle swarm optimizer for mechanical design optimization problems. Engineering Optimization 36(5), 585–605 (2004)
65. Hasançebi, O., Çarbas, S., Dogan, E., Erdal, F., Saka, M.P.: Performance evaluation of metaheuristic search techniques in the optimum design of real size pin jointed structures. Computers and Structures 87(5–6), 284–302 (2009)
66. Kaveh, A., Talatahari, S.: A particle swarm ant colony optimization for truss structures with discrete variables. Journal of Constructional Steel Research (2009)

Determining Viscoelastic and Damage Properties Based on Harmony Search Algorithm

Sungho Mun[1] and Zong Woo Geem[2]

Abstract. This chapter documents the procedure for determining viscoelastic and damage properties using a harmony search (HS) algorithm that employs a heuristic algorithm based on an analogy with music phenomenon. To determine the viscoelastic material parameters, the steps involved in conducting the interconversion between frequency-domain and time-domain functions are outlined, based on the presmoothing of raw data using the HS algorithm. Thus, a Prony series representation of the fitted data can be obtained that includes the determination of the Prony series coefficients. To determine the damage properties of hot mix asphalt (HMA) concrete, a rate-type evolution law is applied for constructing the damage function of the HMA concrete. The damage function can be characterized by fitting experimental results using the HS algorithm. Results from laboratory tests of uniaxial specimens under axial tension at various strain rates are shown to be consistent with the rate-type model of evolution law.

1 Introduction

Linear viscoelastic (LVE) materials are rheological materials that exhibit time-temperature rate-of-loading dependence. When their response is not only a function of the current input, but also of the current and past input history, the characterization of the viscoelastic response can be expressed using the convolution (hereditary) integral. A general overview of time-dependent material properties has been presented [1]. Additionally, a detailed description of the physical response of LVE materials has been explained [2] based on ramp tests to determine the relaxation modulus which is a time-domain LVE response function.

Hot mix asphalt (HMA) concretes used in this study are composite materials consisting of aggregates and asphalt binder. Their behavior is characterized by the interaction between these two components and the LVE behavior of the HMA

[1] Expressway & Transportation Research Institute, Korea Expressway Corporation, Hwaseong, South Korea
Email: smundyna@gmail.com

[2] Environmental Planning and Management Program, Johns Hopkins University, Baltimore, Maryland, USA
Email: zwgeem@gmail.com

Z.W. Geem (Ed.): Harmony Search Algo. for Structural Design Optimization, SCI 239, pp. 199–226.
springerlink.com

concretes, which depends on temperature, loading frequency, and strain magnitude. Studying the behavior of the HMA material requires modeling the LVE behavior through a dynamic modulus test (DMT) conducted in stress-control within the LVE range. This test is run on all previously untested specimens to obtain a LVE fingerprint (linear viscoelastic properties characteristic of HMA specimen) and to determine the shift factors for the undamaged state after constructing dynamic modulus and phase angle mastercurves. Sinusoidal loading in tension and compression sufficient to produce total strain amplitude of about 70 micro-strains was applied at different frequencies. Based on earlier work [3], the 70 micro-strain limit was found not to cause significant damage to the specimen. For mastercurve construction, several replicates are tested at four temperatures: –10, 5, 25 and 40°C.

Several methods have been proposed to convert the dynamic modulus, a LVE response function in the frequency domain, to the corresponding relaxation modulus in the time domain. For a time-domain LVE function, the Prony series is a popular representation, mainly because of its ability to describe a wide range of viscoelastic response and its relatively simple and rugged computational efficiency associated with its exponential basis functions. In this study, an approach is proposed to overcome the problem – namely, oscillations in the fitted curve – associated with the Prony series fitting. The experimental source data are smoothed through a defined log-sigmoidal function using the heuristic optimization technique of the harmony search (HS) algorithm. The coefficients of the log-sigmoidal function can be determined when a defined error norm converges into the minimum point. Thus, this procedure is required to represent the oscillated broadband data using the smoothed log-sigmoidal function. The fitted log-sigmoidal function is used to obtain the time-domain Prony series. This approach has proved very effective and stable in fitting a Prony series with positive coefficients to LVE response function data, as illustrated in an example using experimental data from a dynamic modulus test on HMA concrete.

In recent years, some success has been achieved in developing a mechanistic constitutive equation of HMA concrete for a viscoelastic continuum damage (VECD) model. Kim et al. [4] developed a uniaxial VECD model by applying the elastic-viscoelastic correspondence principle to separate the effects of viscoelasticity, and then employing internal state variables based on the work potential theory to account for the damage evolution under loading. From the verification study it was found that this constitutive model has the ability to predict the hysteretic behavior of the material under both monotonic and cyclic loading up to failure, varying loading rates, random rest durations, multiple stress/strain levels, and different modes of loading (controlled-stress versus controlled-strain). Daniel and Kim [5] discovered a unique damage characteristic curve that describes the reduction in material integrity as damage grows in the HMA specimen, regardless of the applied loading conditions (cyclic versus monotonic, amplitude/rate, and frequency). Chehab et al. [6] demonstrated that the time-temperature superposition principle is valid not only in the LVE state, but also with growing damage. This finding allows the prediction of mixture responses at various temperatures from laboratory testing from a single temperature.

To characterize the damage function with respect to an internal state variable for the uniaxial behavior of an HMA specimen, the HS algorithm can be applied to fit experimental data into a defined damage function, based on minimizing an error norm.

The outline of this chapter is as follows. Section 2 contains the details of the heuristic HS algorithm. Section 3 provides the HMA material parameters, which must be determined by the HS algorithm. The determination of material parameters and their application to the uniaxial behavior of HMA concrete with damage evolution is shown in Section 4. The concluding remarks found in Section 5 summarize the chapter.

2 Harmony Search Algorithm

Many engineering optimization problems are very complex in nature and quite difficult to be solved using gradient-based search algorithms. If there is more than one local optimum in the problem, the result may depend on the selection of an initial point, and the obtained optimal solution may not necessarily be the global optimum. To determine the LVE and damage properties of HMA concrete, the HS algorithm, which has an analogy between music and optimization, was used in this study.

The HS algorithm conceptualizes a behavioral phenomenon of musicians in the improvisation process, where each musician continues to experiment and improve his or her contribution in order to search for a better state of harmony [7, 8]. This section describes the HS algorithm based on the heuristic algorithm that searches for a globally optimized solution. First, a brief overview of the HS algorithm used to formulate solution vectors in which the optimization process is generated and the object function is evaluated, is provided. Finally, the application procedure of the HS algorithm is explained in detail.

2.1 Algorithm Procedure

The procedure for a harmony search, which consists of Steps 1 to 5, is shown in Figure 1. The algorithm parameters are specified in Step 1, as follows: the *harmony memory size* (HMS) is initialized as the number of solution vectors in *harmony memory* (HM); the *harmony memory considering rate* (HMCR, between 0 and 1) is the rate of memory consideration; the *pitch adjusting rate* (PAR, between 0 and 1) is the rate of pitch adjustment; and the maximum number of improvisations, or stopping criterion, is the termination of the HS program. In addition, the optimization problem is specified as follows:

$$\text{Minimize } f(\mathrm{X}) \text{ subject to } x_i \in X_i,\ i = 1,2,\ldots,N\,, \tag{1}$$

where $f(\mathrm{X})$ is an objective function; X is the set of each decision variable, x_i; N is the number of decision variables; X_i is the set of the possible range of values for each decision variable, that is $Lx_i \leq X_i \leq Ux_i$; and Lx_i and Ux_i are the lower and upper bounds for each decision variable, respectively.

The HM is a memory location where all the solution vectors (sets of decision variables) are stored. The HM is similar to the genetic pool in the genetic algorithm (GA) [9]. Here, the HMCR and PAR are parameters that are used to improve the solution vector, as defined in Step 3.

In Step 2, the HM matrix is initially filled with as many randomly generated solution vectors as the HMS, as well as with the corresponding function value of each random vector, $f(X)$.

$$\mathrm{HM} = \left[\begin{array}{cccc|c} x_1^1 & x_2^1 & \cdots & x_N^1 & f(\mathrm{X}^1) \\ x_1^2 & x_2^2 & \cdots & x_N^2 & f(\mathrm{X}^2) \\ \vdots & \vdots & \cdots & \vdots & \vdots \\ x_1^{HMS} & x_2^{HMS} & \cdots & x_N^{HMS} & f(\mathrm{X}^{HMS}) \end{array}\right]. \tag{2}$$

In Step 3, a new harmony vector, $X' = (x_1', x_2', \ldots, x_N')$, is improvised based on the following three mechanisms: 1) random selection, 2) memory consideration, and 3) pitch adjustment. In the random selection, the value of each decision variable, x_i', in the new harmony vector is randomly chosen within the value range with a probability of (1 – HMCR). The HMCR, which varies between 0 and 1, is the rate of choosing one value from the historical values stored in the HM, and (1 – HMCR) is the rate of randomly selecting one value from the possible range of values.

The value of each decision variable obtained by the memory consideration is examined to determine whether it should be pitch-adjusted. This operation uses the PAR parameter, which is the rate of pitch adjustment as it should be pitch-adjusted to neighboring pitches with a probability of HMCR × PAR, while the original pitch obtained in the memory consideration is kept with a probability of HMCR × (1 – PAR). For example, this operation uses the PAR parameter, which ranges between 0 and 1.

If the pitch adjustment decision for x_i' is made with a probability of PAR, x_i' is replaced with $x_i' \pm rand \times bw$, where *rand* and *bw* are random numbers (e.g., a value between 0 and 1) and bandwidths between the lower and upper bounds, respectively. The value of (1 – PAR) sets the rate of performing nothing. Thus, the pitch adjustment is applied to each variable as follows:

$$x_i' \leftarrow \begin{cases} x_i' + rand \times bw, \text{ with a probability of } HMCR \times PAR \times 0.5 \\ x_i' - rand \times bw, \text{ with a probability of } HMCR \times PAR \times 0.5 \\ x_i', \text{ with a probability of } HMCR \times (1 - PAR) \end{cases} \tag{3}$$

If the new harmony vector is better than the worst harmony in the HM, based on the evaluation of an objective function value, the new harmony is included in the HM, and the existing worst harmony is excluded from the HM. Finally, if the stopping criterion (or maximum number of improvisations) is satisfied, the computation is terminated. Otherwise, Steps 3 and 4 are repeated.

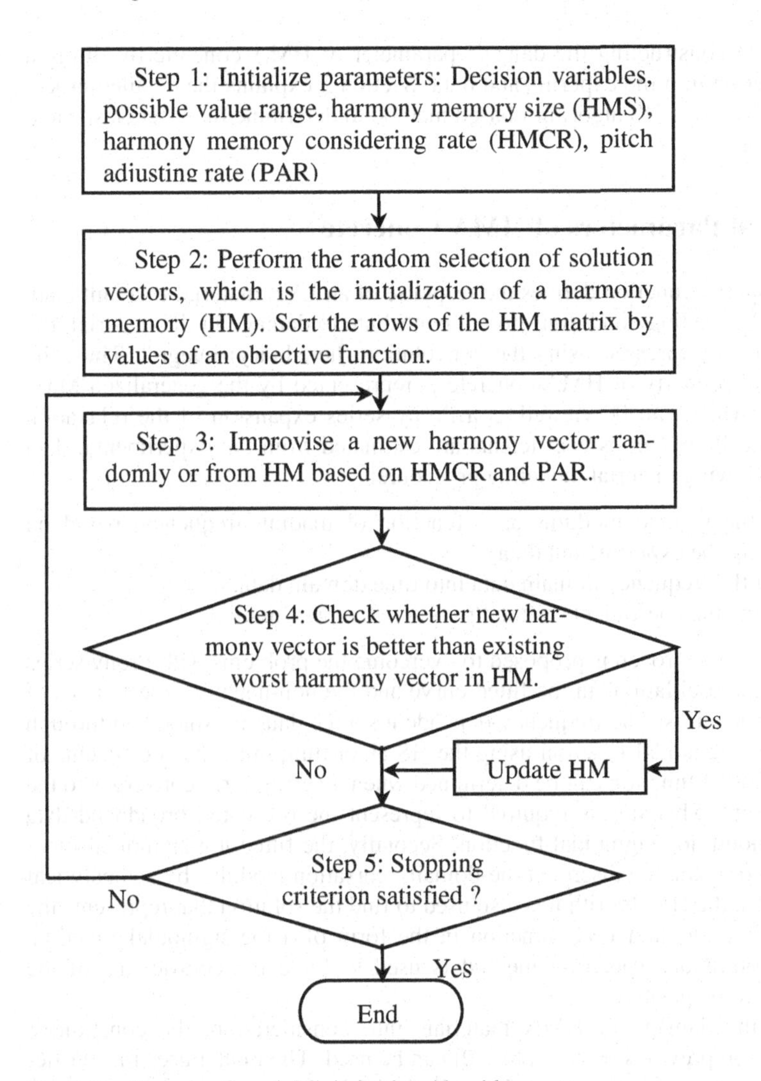

Fig. 1 The Harmony Search Minimization Algorithm

2.2 *Application to the Determination of Viscoelastic and Damage Properties of HMA Concrete*

In terms of applying the HS algorithm to the determination of viscoelastic and damage properties of HMA materials, a VECD model is used, which is generally applied to describe the fatigue phenomenon occurring at the HMA layers of a flexible pavement. Thus, the optimization problems, which are related to the VECD model, can be solved using the HS algorithm process with respect to a) obtaining the LVE properties from experimental data, b) determining Prony series parameters that are used to represent the LVE HMA by a generalized Maxwell model, and c) constructing the damage parameter of HMA concrete by fitting a defined function into the experimental data. Section 3 explains the detailed procedure of the proposed HS algorithm-based method to determine the viscoelastic and damage properties.

3 Material Parameters of HMA Concrete

An LVE response function defines the response of an LVE material to a unit load. As long as the loading conditions do not contribute to damage in the material, the response can be expressed using the convolution (hereditary) integral. Thus, the LVE material property of HMA concrete is represented by the generalized Maxwell model, which can be viewed as a Prony series expansion of the relaxation modulus. The Prony series coefficients are estimated from the experimental data using the following material modeling procedure:

1) Obtain the storage modulus as a function of loading frequency, based on smoothing the experimental data.
2) Convert the frequency domain data into time domain data.
3) Determine the coefficients of the Prony series representation.

In this study, an approach is proposed to overcome the problems with Prony series fitting, i.e., the oscillations in the fitted curve and the non-negative coefficients of the Prony series. First, the frequency-dependent source data are smoothed through a defined log-sigmoidal function using the HS algorithm; thus, the coefficients of the log-sigmoidal function can be determined when an error norm converges to the minimum point. This step is required to represent the oscillated broadband data using the smooth log-sigmoidal function. Secondly, the fitted log-sigmoidal function is used to obtain the discrete time-domain relaxation modulus by an analytical method. Then, the HS algorithm is also used to find the solution that represents the continuous time-domain LVE function in the form of a log-sigmoidal function. Finally, a continuous spectrum method is used to force the coefficients of the Prony series to be positive.

To take the damage of HMA materials into consideration, the constitutive model based on previous research [4, 10] can be used. The model uses the elastic-viscoelastic correspondence principle to eliminate the time dependence of the material. The work potential theory [11, 12] is then used to model the damage growth in the material as well as to include the viscoelastic effects of microcracking. The work potential theory proposes the following rate-type damage evolution law:

$$\dot{S}_m = \left(-\frac{\partial W^R}{\partial S_m} \right)^{\alpha_m}, \tag{4}$$

where the overdot represents the derivative with respect to time; W^R is the pseudo strain energy density function; α_m is a material-dependent constant; and m is not summation.

In its application to the uniaxial behavior of HMA concrete [13], the experimental stress-strain constitutive relationship is incorporated into the one-dimensional pseudo strain energy density function of the material in the following form:

$$W^R = \frac{1}{2} C(S)(\varepsilon^R)^2, \quad (5)$$

where the damage function, $C(S)$, depends on a single damage parameter, S; and ε^R is the pseudo strain. In order to determine a state variable, S, which is used to trace a state of material damage according to the relationship with the pseudo strain energy density function in Equation (5), the HS algorithm can be applied for fitting the experimental data into a defined function of $C(S)$ based on minimizing an error norm.

3.1 Testing Systems and Methods

In this study, 12.5 mm and 9.5 mm mixtures were used for unmodified HMA and lime-modified HMA, respectively. The 12.5 mm and 9.5 mm mixtures were used based on the Superpave mix design [14]. The aggregate blend used consisted of 95.5% by mass granite aggregates, 3.5% natural sand, and 1% baghouse fines. In case of the lime-modified HMA, hydrated lime (1% by aggregate weight) is substituted for a portion of the baghouse fines. The asphalt binders used were performance grade (PG) 70-22 and PG 58-28 for unmodified and lime-modified HMA, respectively. The asphalt contents were 5.2% for the unmodified HMA; 5.3% and 5.8% for the lime-modified HMA by mass. Mixing and compaction temperatures were 166°C and 153°C, respectively. Compaction was done using the Superpave gyratory compactor.

The mixtures were compacted into gyratory plugs of 150 mm in diameter by 178 mm in height. Then, they were cut and cored to cylindrical specimens with dimensions of 100 mm in diameter and 150 mm in height for dynamic modulus tests and 75 mm in diameter and 140 mm in height for constant crosshead tests.

Regarding testing setup and methods, two different closed-loop servo-hydraulic testing machines were utilized in this study. The first is a MTS 810 loading frame equipped with either a 25 kN or 8.9kN load cell, depending on the nature of the test. For this machine, an environmental chamber, equipped with liquid nitrogen coolant and a feedback system, was used to control and maintain the test temperature. The second machine is a UTM-25 machine, manufactured by IPC Global of Australia. This machine is equipped with a 25 kN load cell. The environmental chamber for the UTM-25 is refrigerator-driven and also uses a feedback system to maintain a consistent temperature during the testing.

Measurements of axial deformations, load, and crosshead movements were taken for all tests. The data acquisition system of both machines was identical, consisting of a National Instruments 16-bit data acquisition card and Labview software. In all tests conducted in this study, axial displacement measurements were taken with linear variable differential transformers (LVDTs) from IPC Global.

Fig. 2 LVDTs Mounting on Specimens: (a) Dynamic Modulus Test; and (b) Constant Crosshead Rate Test

Tests conducted in this study include dynamic modulus testing and constant crosshead rate testing. All tests were done in both machines. The dynamic modulus test (DMT) shown in Figure 2a is performed by applying a sinusoidal load to an asphalt concrete specimen to obtain the linear viscoelastic material properties of asphalt mixtures. The loading amplitude is adjusted based on the material stiffness, temperature, and frequency to keep the strain response within the linear viscoelastic range.

Regarding the determination of the damage function, $C(S)$, the constant crosshead rate tests were conducted in tension mode till failure of the specimen at different crosshead rates as shown in Figure 2b. Testing temperatures were 5°C and 25°C for the lime-modified HMA and the unmodified HMA, respectively, in this study.

3.2 Determination of LVE Material Parameters

In order to get the experimental data to be fitted in this study, a DMT is performed by controlling a micro-strain level of 70 that is targeted as the limit for the LVE. The loading is applied until steady-state response is achieved, at which point several cycles of data are collected. After each frequency, a five-minute rest period is allowed for specimen recovery before the next loading block is applied. The frequencies are applied from the fastest to the slowest ranging from 1 to 20 Hz.

From the DMT, the complex modulus, E^*, which includes the dynamic modulus ($|E^*|$) and the phase angle (ϕ), can be determined. The complex modulus can also be viewed as a composition of storage (E') and loss modulus (E'') as follows:

$$E^* = E' + iE'' \tag{6}$$

where i is the $\sqrt{-1}$. The dynamic modulus is the amplitude of the complex modulus and is defined as:

$$|E^*| = \sqrt{(E')^2 + (E'')^2}\,. \tag{7}$$

The values of the storage and loss modulus, which are shown in Figure 3, are related to the dynamic modulus and phase angle as follows:

$$E' = |E^*|\cos\phi \;\text{ and }\; E'' = |E^*|\sin\phi\,. \tag{8}$$

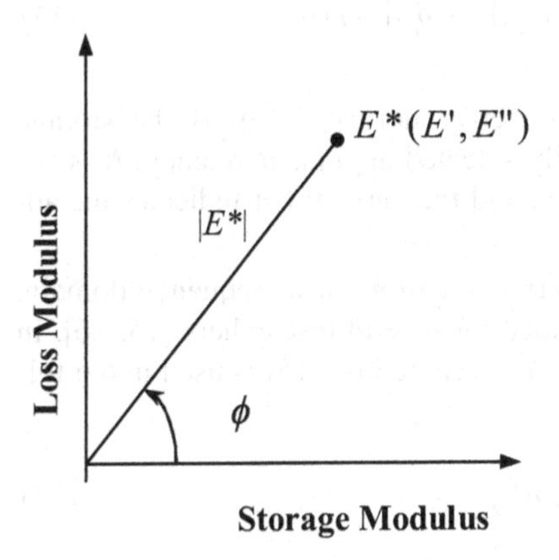

Fig. 3 Complex Modulus Schematic Diagram

As the material becomes more viscous, the phase angle increases and the loss component of the complex modulus increases. Conversely, a decreasing phase angle indicates more elastic behavior and a larger contribution from the storage modulus. The dynamic modulus at each frequency is calculated by dividing the steady state stress amplitude, σ_{amp}, by the strain amplitude, ε_{amp}:

$$|E^*| = \frac{\sigma_{amp}}{\varepsilon_{amp}}\,. \tag{9}$$

The phase angle, ϕ, is associated with the time lag, Δt, between the strain input and stress response at the corresponding frequency, f:

$$\phi = 2\pi f \Delta t\,. \tag{10}$$

In order to determine the storage modulus prior to the conversion (i.e., from frequency domain to time domain) using a Prony series function, the discrete raw data need to be fitted using a continuous function. Thus, the quality of raw data can be significantly improved using a defined log-sigmoidal function that describes the full viscoelastic range of the HMA, from the glassy state to the low frequency plateau. The log-sigmoidal function, $f(\omega)$, is defined as

$$f(\omega) = a_1 + \frac{a_2}{\left\{a_3 + \frac{a_4}{\exp\left[a_5 + a_6 \log_{10}(\omega)\right]}\right\}} \tag{11}$$

where $a_{1,2,\ldots,6}$ are the coefficients determined by the HS algorithm, and ω is the angular frequency. For example, the optimizing solution between the storage modulus data and the log-sigmoidal function that is used to determine the coefficients of Equation (11) can be expressed by the following Equation (12):

$$\text{Minimize } g(\omega_i) = \sum_{i=1}^{N} \left|\log_{10}\left[E'(\omega_i)\right] - f(\omega_i)\right|, \tag{12}$$

where $g(\omega_i)$ is the objective function (e.g., error norm); $E'(\omega_i)$ is the storage modulus; subscript i denotes the individually selected angular frequency; N is the total number of selected angular frequencies; and the vertical bar indicates the absolute value.

The interconversion between LVE material functions of a frequency-domain, E', and a time-domain, $E(t)$, has been provided by several researchers [15, 16]. In this study, an approximate analytical solution developed by [15] is used in the following form:

$$E(t) \cong \frac{1}{\lambda'} E'(\omega)\,|_{\omega=(1/t)}, \tag{13}$$

where λ' is an adjustment factor that is defined by $\Gamma(1\text{-}n)\cos(n\pi/2)$; Γ is a gamma function; and n is the local log-log slope of the storage modulus, that is,

$$n = \frac{d \log_{10} E'(\omega)}{d \log_{10} \omega}. \tag{14}$$

Thus, the time-domain relaxation modulus can also be fitted using the defined log-sigmoidal function with respect to time, which is the same as Equation (11), based on the above function-fitting algorithm.

In order to introduce the time-domain Prony series representation, the uniaxial, non-aging, isothermal stress-strain constitutive equation for a LVE material can be considered, as follows:

$$\sigma(t) = \int_0^t E(t-\tau) \frac{d\varepsilon(\tau)}{d\tau} d\tau, \tag{15}$$

where $\sigma(t)$, $\varepsilon(t)$, and $E(t)$ are the stress, strain, relaxation modulus components in a time domain t, respectively. The relaxation modulus of Equation (13), $E(t)$, which is based on a generalized Maxwell model consisting of a series of springs and dashpots [17] in the form of a Prony series, can be expressed as follows:

$$E(t) = E_\infty + \sum_{m=1}^{M} E_m \exp(-t/\rho_m), \tag{16}$$

where E_∞, ρ_m, and E_m are the infinite relaxation modulus, relaxation time, and Prony coefficients, respectively.

Considering the case of an infinite number of Maxwell components with continuously distributed relaxation times, and neglecting the infinite relaxation modulus, E_∞, based on the continuous spectrum method [18-20], the relaxation modulus can be defined by

$$E(t) = \int_{-\infty}^{\infty} L(\rho)\exp(-t/\rho)d\rho\,, \tag{17}$$

where L(ρ) represents a continuous distribution of the relaxation modulus, and is defined by

$$L(\rho) = \lim_{k\to\infty}\frac{(-k\rho)^k}{(k-1)!}E^{(k)}(k\rho)\,. \tag{18}$$

In order to determine the discrete Prony series coefficients, E_m, in Equation (16) from the continuous spectrum equation of Equation (17), the continuous spectrum can be approximated by subdividing lnρ into time intervals $\Delta(\ln\rho_m)$ = $\ln 10\Delta(\log_{10}\rho_m)$, as follows:

$$E(t) = \int_{-\infty}^{\infty} L(\rho)\exp(-t/\rho)d\ln\rho \approx \sum_{m=1}^{M} L(\rho_m)\exp(-t/\rho_m)\Delta(\ln\rho_m)\,. \tag{19}$$

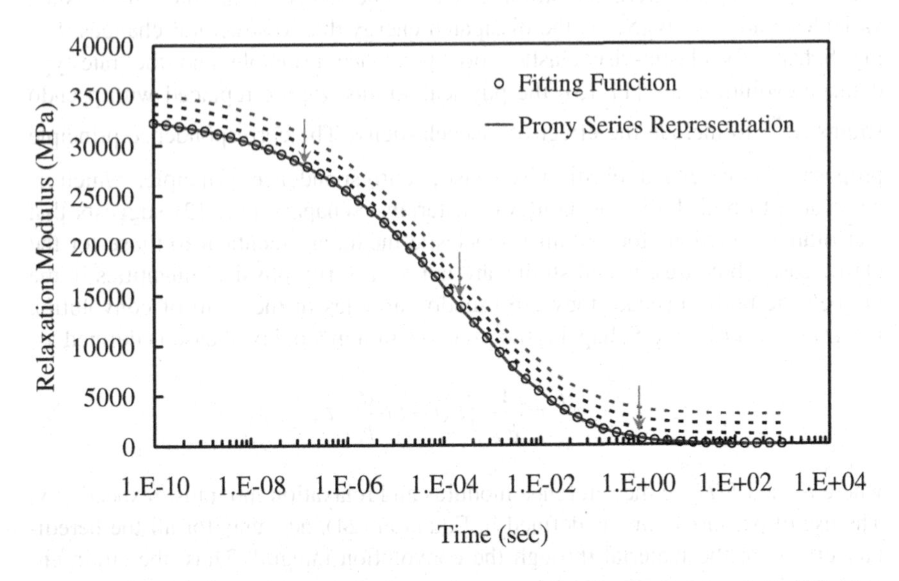

Fig. 4 Adjusting the Vertical Shift of a Relaxation Modulus Curve through the HS Algorithm to Obtain an Infinite Relaxation Modulus, E_∞

The Prony coefficients for the chosen relaxation time can be determined by

$$E_m = L(\rho_m)\ln 10\Delta(\log_{10}\rho_m). \quad (20)$$

Finally, the infinite relaxation modulus, E_∞, of a Prony series representation can be found by adjusting the vertical shift of a relaxation modulus curve through the HS algorithm, as shown in Figure 4.

3.3 Determination of Damage Function, C(S)

Schapery [12] developed a theory using the method of thermodynamics of irreversible processes to describe the mechanical behavior of elastic composite materials with growing damage. Three fundamental elements comprise the work potential theory:

1) Strain energy density function, $W = W(\varepsilon_{ij}, S_m)$ (21)

2) Stress-strain relationship, $\sigma_{ij} = \dfrac{\partial W}{\partial \varepsilon_{ij}}$ (22)

3) Damage evolution law, $-\dfrac{\partial W}{\partial S_m} = \dfrac{\partial W_s}{\partial S_m}$, (23)

where σ_{ij} and ε_{ij} are stress and strain tensors, respectively; S_m are the internal state variables; and $W_s = W_s(S_m)$ is the dissipated energy due to structural changes. Using Schapery's elastic-viscoelastic correspondence principle and the rate-type damage evolution law [11-13], the physical strains, ε_{ij}, are replaced with pseudo strains, ε_{ij}^R, to include the effect of viscoelasticity. The correspondence principle proposes the extended elastic-viscoelastic correspondence principle, which is applicable to both LVE and nonLVE materials. Schapery [11, 12] suggests that constitutive equations for certain viscoelastic media are identical to those for the elastic cases, but stresses and strains are not necessarily physical quantities in the viscoelastic body. Instead, they are pseudo variables in the form of convolution integrals. According to Schapery, the pseudo strain in a uniaxial case is defined as

$$\varepsilon^R = \frac{1}{E_R}\int_0^t E(t-\tau)\frac{d\varepsilon}{d\tau}d\tau, \quad (24)$$

where E_R and $E(t)$ are the reference modulus and relaxation modulus, respectively. The use of pseudo strain, as defined in Equation (24), accounts for all the hereditary effects of the material through the convolution integral. Thus, the strain energy density function, $W = W(\varepsilon, S)$, transforms to the pseudo strain energy density function.

The pseudo strain energy density function of the material is formulated using Equation (22). For the uniaxial case, the stress can be determined by using the relationship of Equation (5), as follows:

$$\sigma = \frac{\partial W^R}{\partial \varepsilon^R} = C(S)\varepsilon^R. \tag{25}$$

The damage evolution law, Equation (4), is reduced to the following single equation for S:

$$\dot{S} = \left(-\frac{\partial W^R}{\partial S} \right)^{\alpha}. \tag{26}$$

To characterize the function $C(S)$ in Equation (25), the damage evolution law and experimental data are used. Using the measured stresses and calculated pseudo strains, the C values can be determined through Equations (24) and (25). For uniaxial loading conditions, a single damage variable, S, is used along with the associated power, α. Using the experimental data, the following incremental relationship can be obtained by combining Equations (5) and (26):

$$\Delta S = \left[\left\{ -\frac{1}{2} \Delta C (\varepsilon^R)^2 \right\}^{\alpha} \Delta t \right]^{1/(1+\alpha)} \quad \text{and} \tag{27}$$

$$S \cong \sum_{n=1}^{N} \left[\frac{1}{2} (C_{n-1} - C_n)(\varepsilon_n^R)^2 \right]^{\alpha/(1+\alpha)} (t_n - t_{n-1})^{1/(1+\alpha)}. \tag{28}$$

The value of α can be found by using the following relationship: $\alpha = 1 + 1/n$, in which n = - log $E(t)$/log(t). Furthermore, a relationship is constructed between C and S, based on a functional form, as follows:

$$C(S) = \exp(-\beta \cdot S^{\gamma}), \tag{29}$$

where β and γ are the parameters that are determined through the best fitting process of the HS algorithm between the functional form and the experimental data.

4 Determination of Material Parameters and Uniaxial Behaviour of HMA Concrete with Damage Evolution

A detailed application of the HS algorithm to determine the material parameters is as follows: 1) determine the coefficients of the fitting function described in Equation (11); 2) determine the Prony series coefficients based on the fitting function; 3) characterize the damage function, $C(S)$, using experimental data; and 4) predict the damage behavior of HMA concrete at various strain input rates.

4.1 Determination of the Coefficients of the Fitting Function

In order to smooth the discrete raw data that need to be fitted using the defined log-sigmoidal function of Equation (11), the HS algorithm can be used based on minimizing the objective function of Equation (12). Figure 5 shows the error norm of the objective function defined by Equation (12) in terms of the number of iterations. Based on Figure 5, it is noted that the best solution converges as the number of iterations increases. Finally, Figure 6 shows the log-sigmoidal function smoothly fitted with the raw data; therefore, the coefficients of Equation (11) in the frequency domain can be found through the HS algorithm as shown in Table 1.

At the low frequencies shown in Figure 6b, some minor loss of information may be resulted due to some local irregularities of the scattered data. However, the fitting function provides a smooth representation over all frequencies.

4.2 Time-Domain Prony Series Representation

To determine the Prony series coefficients, the analytical solution of Equation (13) is utilized to convert a frequency-domain to a time-domain modulus; also, the time-domain modulus is fitted with the defined log-sigmoidal function of Equation (11) as shown in Table 1. Based on Equation (20), the Prony coefficients for the chosen relaxation times can be found, and the infinite relaxation modulus can be found, as shown in Figure 4. Finally, Prony series representation of the relaxation modulus is shown in Figure 7. Table 2 shows the Prony series coefficients of the unmodified and lime-modified HMA concretes.

4.3 Comparison between HS Algorithm and Regression Method in Terms of Prony Series Representation

The regression method, which is used for comparing with the HS algorithm, is based on fitting raw data into a predefined function through a least squares method as shown in Figure 8. Thus, the storage modulus is expressed by the regression equation that is similar to other work [4].

In order to determine the Prony series coefficients in the regression model, a viscoelastic constitutive relationship derived from the generalized Maxwell model can be used. The mechanical model consists of a series of springs and dashpots in an arrangement shown in Figure 9. For a given applied strain, ε, the stress in the single spring, σ_∞, is given as follows:

$$\sigma_\infty = E_\infty \varepsilon\ . \tag{30}$$

The stress, σ_m, in each of the Maxwell components combining a spring with a dashpot is governed by the differential equation:

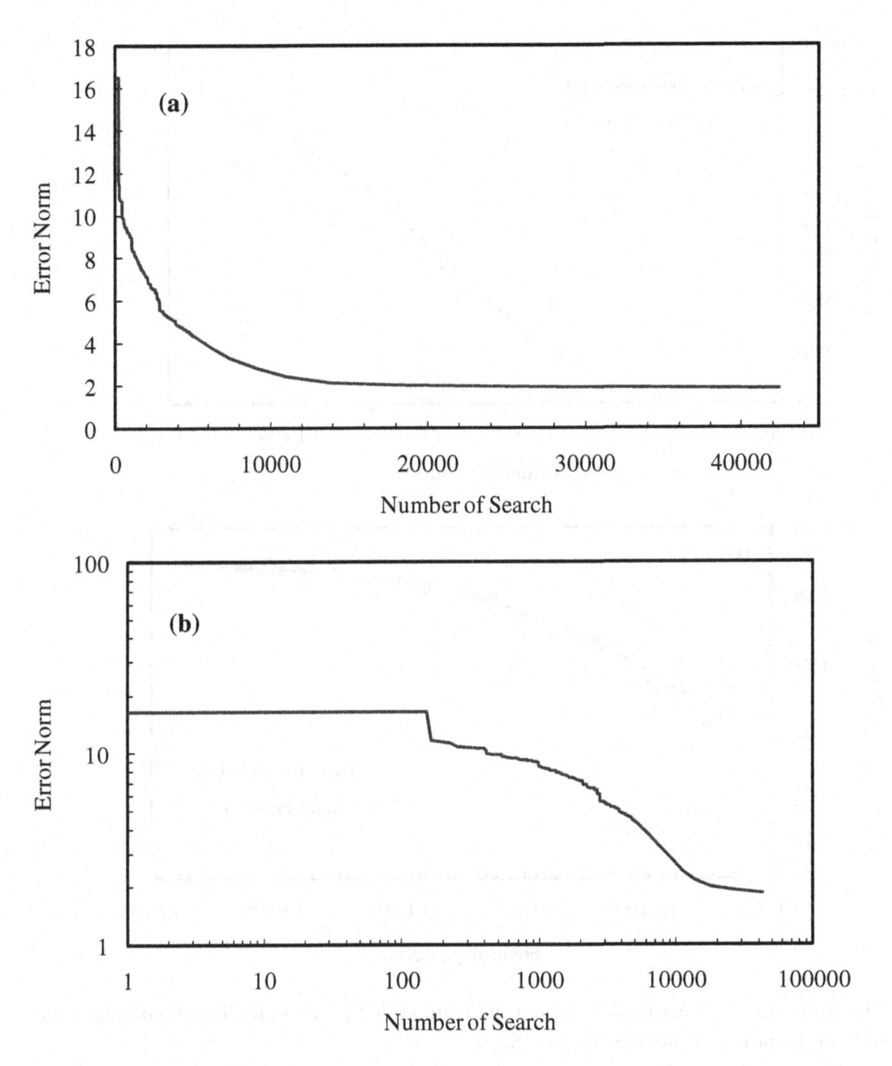

Fig. 5 Convergence of the Objective Function during the Number of Searches: (a) Semi-Log Scale; (b) Log-Log Scale

$$\frac{d\varepsilon}{dt} = \frac{1}{E_m}\frac{d\sigma_m}{dt} + \frac{\sigma_m}{\eta_m} \tag{31}$$

where η_m is the coefficient of viscosity, and E_m is the spring stiffness (e.g., Prony coefficient) in the mth term or Prony coefficient. Based on the linearity of the material components, the total stress in the generalized Maxwell model is obtained by a summation as follows:

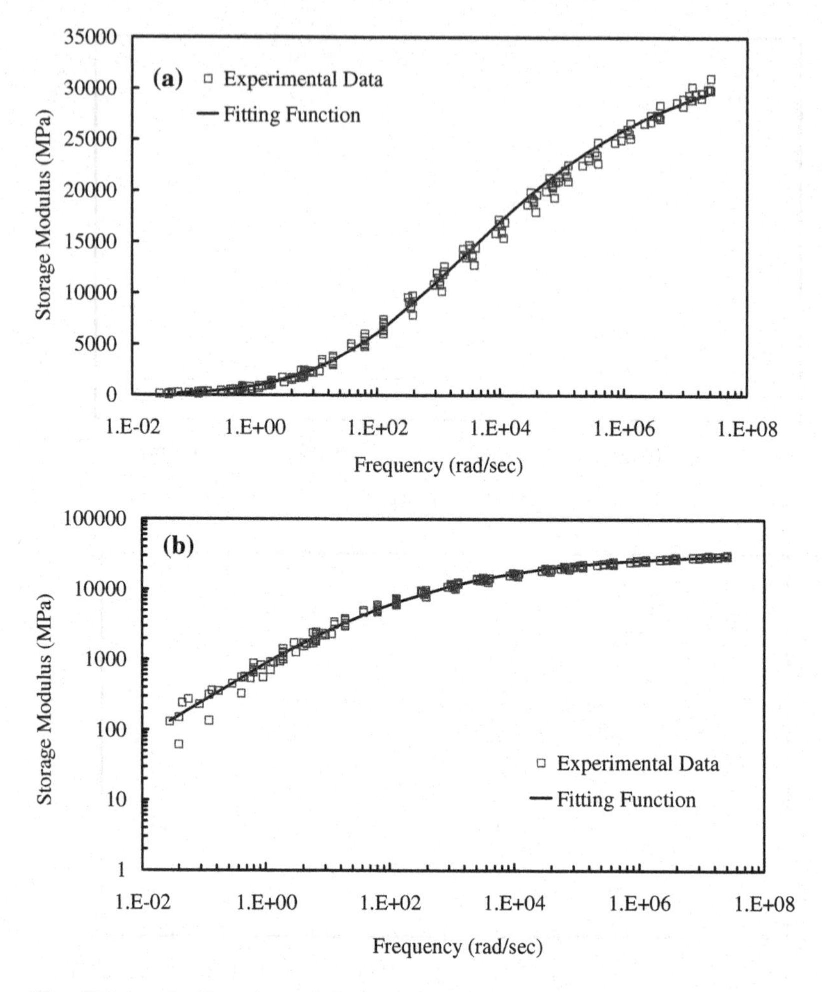

Fig. 6 Fitting the Experimental Data of the Unmodified HMA to the Log-Sigmoidal Function: (a) Semi-Log Scale; (b) Log-Log Scale

$$\sigma = \sigma_\infty + \sum_{m=1}^{M} \sigma_m \, . \tag{32}$$

Fourier transform is useful in solving the above differential equation based on the elastic-viscoelastic correspondence principle, where elastic moduli are replaced by their corresponding viscoelastic counterparts in the Fourier transform domain [21, 22]. Hence, the differential equation relating stress to strain is converted into an algebraic equation. Applying the Fourier-transform technique to Equations (30) to (32), and then eliminating the stresses σ_∞ and σ_m from the equation gives:

Table 1 Coefficients of the Log-Sigmoidal Function in the Time and Frequency Domain

		a_1	a_2	a_3	a_4	a_5	a_6
Unmodified HMA (Binder content: 5.2%)	Coefficients in frequency	4.51792	1.21033	-0.29528	-4.11821	2.17693	-0.52211
	Coefficients in time	4.52263	-2.46005	0.59640	1.28745	0.39382	0.52051
Lime-modified HMA (Binder content: 5.3%)	Coefficients in frequency	1.78212	-5.38729	-1.92668	-0.40234	-0.17790	0.48422
	Coefficients in time	1.59498	-5.32351	-1.76661	-0.69466	0.34174	-0.43839
Lime-modified HMA (Binder content: 5.6%)	Coefficients in frequency	2.02698	-1.51221	-0.59659	-5.22501	3.28673	0.49089
	Coefficients in time	1.94225	13.47153	5.11108	0.62004	-1.09828	-0.46373

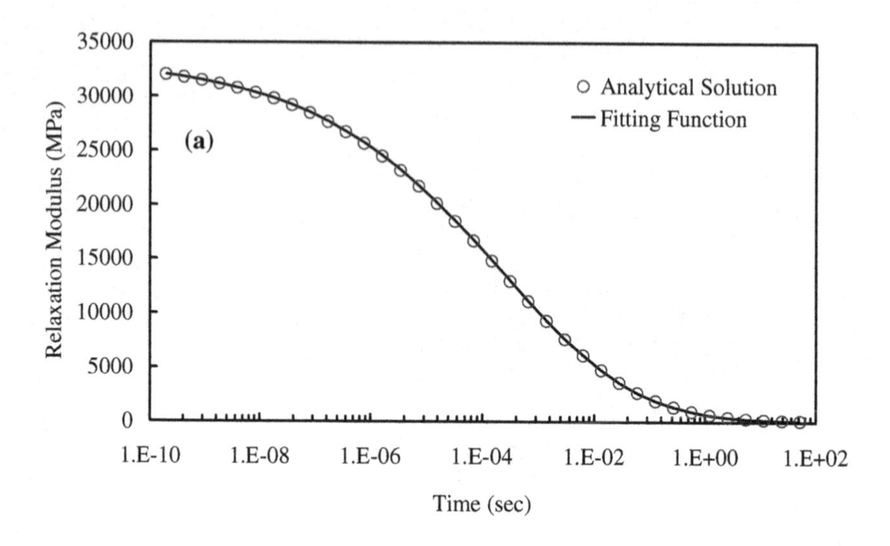

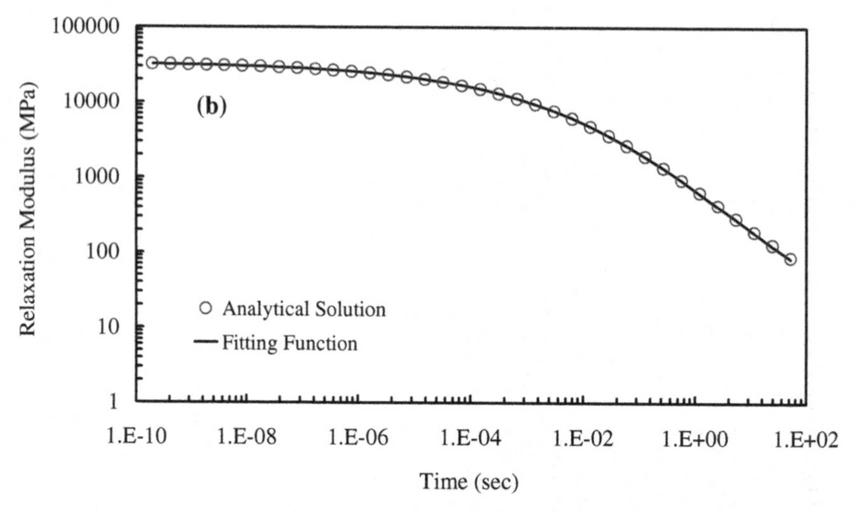

Fig. 7 Fitting the Experimental Data of the Unmodified HMA to the Log-Sigmoidal Function: (a) Semi-Log Scale; (b) Log-Log Scale

$$\breve{\sigma} = \left(E_{\infty} + \sum_{m=1}^{M} \frac{i\omega_n \rho_m E_m}{i\omega_n \rho_m + 1} \right) \breve{\varepsilon}\ ,\ n=1,\ldots,N \tag{33}$$

where $\breve{\sigma}$ and $\breve{\varepsilon}$ are defined as stress and strain in the Fourier-transform domain, with the relaxation time of the mth Maxwell element expressed as follows:

$$\rho_m \equiv \frac{\eta_m}{E_m}\ . \tag{34}$$

Table 2 Prony Series Coefficients of the Unmodified and Lime-Modified HMA Concretes

Unmodified HMA (Binder content: 5.2%)		Lime-modified HMA (Binder content: 5.3%)		Lime-modified HMA (Binder content: 5.6%)	
E_∞: 22.40 MPa		E_∞: 879.58 MPa		E_∞: 800.46 MPa	
ρ_m	E_m (MPa)	ρ_m	E_m (MPa)	ρ_m	E_m (MPa)
1.E-10	507.22	1.E-10	366.30	1.E-10	317.03
1.E-09	834.93	1.E-09	560.07	1.E-09	497.47
1.E-08	1353.67	1.E-08	849.91	1.E-08	774.59
1.E-07	2139.18	1.E-07	1274.78	1.E-07	1191.33
1.E-06	3237.37	1.E-06	1877.92	1.E-06	1796.96
1.E-05	4555.73	1.E-05	2690.98	1.E-05	2628.59
1.E-04	5683.89	1.E-04	3696.78	1.E-04	3665.22
1.E-03	5855.13	1.E-03	4766.10	1.E-03	4748.17
1.E-02	4555.50	1.E-02	5597.54	1.E-02	5513.79
1.E-01	2485.68	1.E-01	5765.33	1.E-01	5492.53
1.E+00	949.99	1.E+00	4999.45	1.E+00	4501.30
1.E+01	282.39	1.E+01	3541.64	1.E+01	2976.67
1.E+02	77.29	1.E+02	2047.75	1.E+02	1624.62
1.E+03	22.60	1.E+03	1005.01	1.E+03	780.13

The complex modulus can be obtained from the constitutive equation shown in Equation (33) according to the following equation:

$$E^* = E_\infty + \sum_{m=1}^{M} \frac{i\omega_n \rho_m E_m}{i\omega_n \rho_m + 1}, \; n=1,\ldots,N. \tag{35}$$

From Equation (35), the storage modulus in frequency-domain can be determined by taking the real parts of the complex modulus:

$$E'(\omega_n) = E_\infty + \sum_{m=1}^{M} \frac{\omega_n^{\;2} \rho_m^2 E_m}{\omega_n^{\;2} \rho_m^2 + 1}, \; n=1,\ldots,N. \tag{36}$$

In order to obtain the time-domain relaxation modulus, the Prony series function in Equation (16) is determined by using the equivalent E_∞, ρ_m, and E_m shown in Equation (36); E_∞ can be found by the limit of $E'(\omega)|_{0<\omega<<1}$. The Prony-series coefficients, E_m, are obtained based on the selected relaxation times and reduced frequencies, ρ_m and ω_n, and the following linear algebraic equation:

$$\vec{f} = \mathbf{E}^{-1}\vec{d} \tag{37}$$

where the column vectors, $\vec{f}$ and $\vec{d}$, are E_m and $E'(\omega_n)$-E_∞ respectively; the superscript –1 denotes an inversion; the matrix, $\mathbf{E}^{-1}$, is as follows:

$$\mathbf{E} = E_{n,m} = \sum_{m=1}^{M} \frac{\omega_n^{\ 2} \rho_m^2}{\omega_n^{\ 2} \rho_m^2 + 1}, \ n=1,\ldots,N. \tag{38}$$

Based on the HS algorithm and regression method, two Prony series representations of relaxation modulus are shown in Figure 10 using the unmodified HMA. In case of the regression method, some oscillation can be observed between 0.01 and 1000 sec. Furthermore, some Prony coefficients shown in Table 3 resulted in negative values, which do not physically make sense because spring stiffnesses should be positive. In order to obtain the positive values of moduli in case of the regression approach, a non-linear optimization process [23] is necessary to approach a better solution.

However, the HS algorithm used in this study addresses the problem of negative Prony series coefficients or oscillations based on fitting raw data with the sigmoidal function when the source data exhibit significant variability.

4.4 Determination of the Parameters of Damage Function C(S)

In order to calculate the damage parameter, S, in terms of the damage function, $C(S)$, the value of α is found to be equal to 2.9 using the log-scale slope, n, in case of the unmodified HMA concrete. Using the HS algorithm as well as the raw data obtained from the calculation of Equation (28), the parameters β and γ of Equation

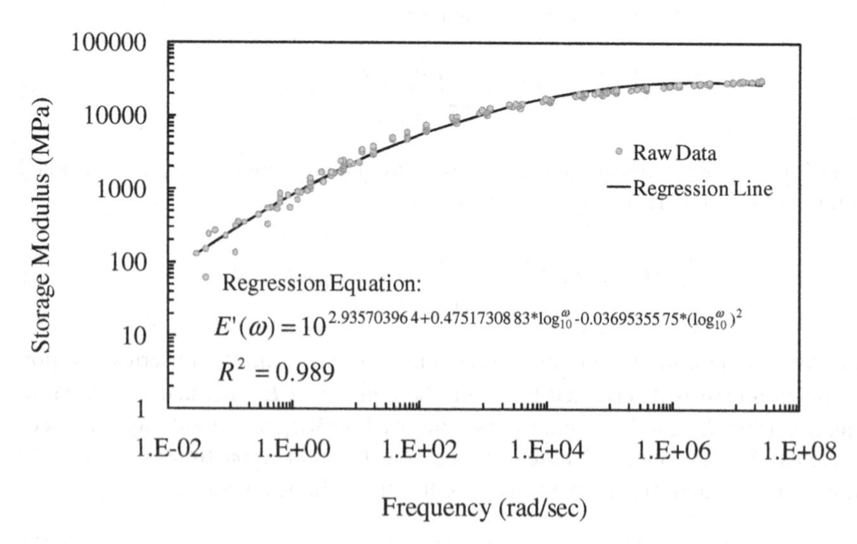

Fig. 8 Fitting Raw Data into a Predefined Function through a Least Squares Method

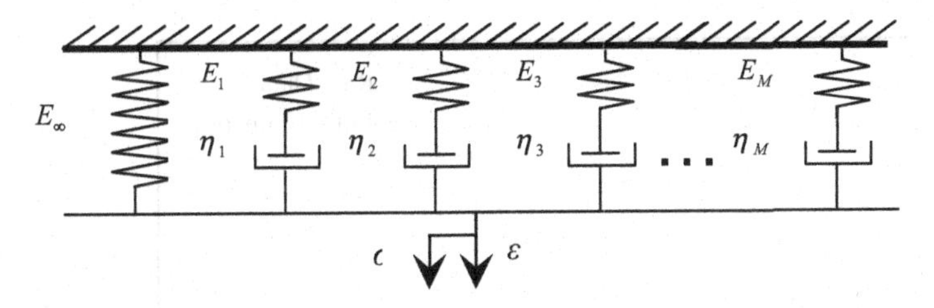

Fig. 9 Generalized Maxwell Model Used for the LVE Modeling

(29) are determined as 0.002757 and 0.543455, respectively, in case of the unmodified HMA concrete. Figure 11 shows the defined function that is fitted using the experimental data of the unmodified HMA. As the same of the determination procedure of the unmodified HMA material parameters, the material parameters of the lime-modified HMA were found as shown in Table 4.

4.5 Application to the Prediction of the Damage Behaviour of HMA Concrete

The material parameters, which are determined by the HS algorithm presented in Section 3, are used for the verification using experimental results. A series of uniaxial extension tests was performed on the cylindrical specimens at 25°C at different constant strain rates, 0.0063/sec and 0.0294/sec, for the unmodified HMA; at 5°C at different constant strain rates, 0.00003/sec and 0.000055/sec, for the lime-modified HMA. Each specimen was glued to the end plates, which were then

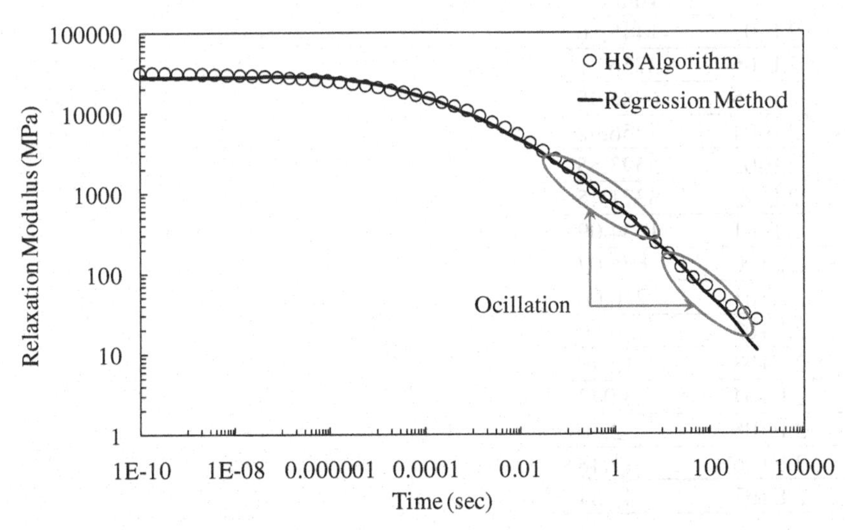

Fig. 10 Prony Series Representations Based on the HS Algorithm and Regression Method

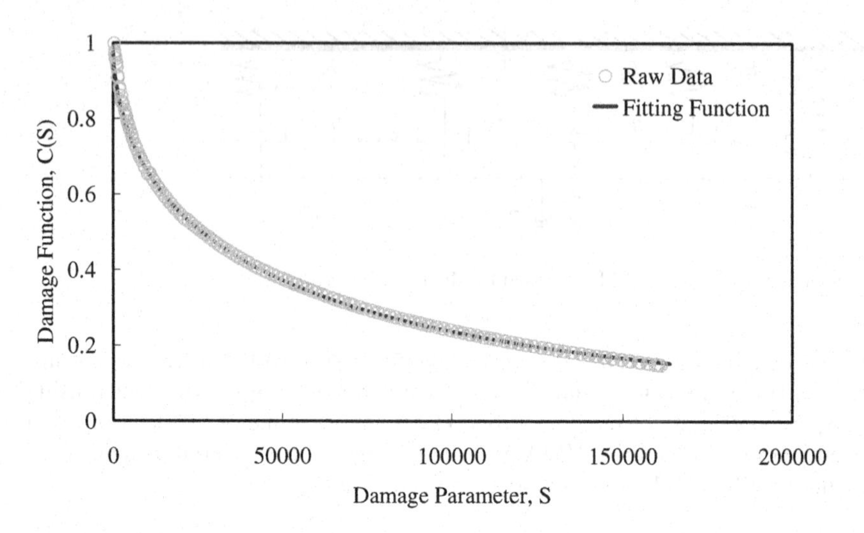

Fig. 11 Damage Function/Parameter Curve of the Unmodified HMA for Raw Data and Fitting Function

Table 3 Prony Coefficients Obtained by Using the Regression Model

Unmodified HMA (Binder content: 5.2%)	
E_∞: 34.40 MPa	
ρ_m	E_m (MPa)
2.E-07	-1445.86
2.E-06	3756.61
2.E-05	6886.95
2.E-04	7256.92
2.E-03	5593.55
2.E-02	3371.45
2.E-01	1634.09
2.E+00	646.07
2.E+01	210.06
2.E+02	56.50
2.E+03	12.40
2.E+04	3.043
2.E+05	-2.50
2.E+06	11.48
2.E+07	-43.54

Table 4 Material Parameter of the Lime-Modified HMA

Material parameters	Lime-modified HMA (Binder content: 5.3%)	Lime-modified HMA (Binder content: 5.6%)
α	2.79	3.06
β	0.001247	0.539442
γ	0.002001	0.503317

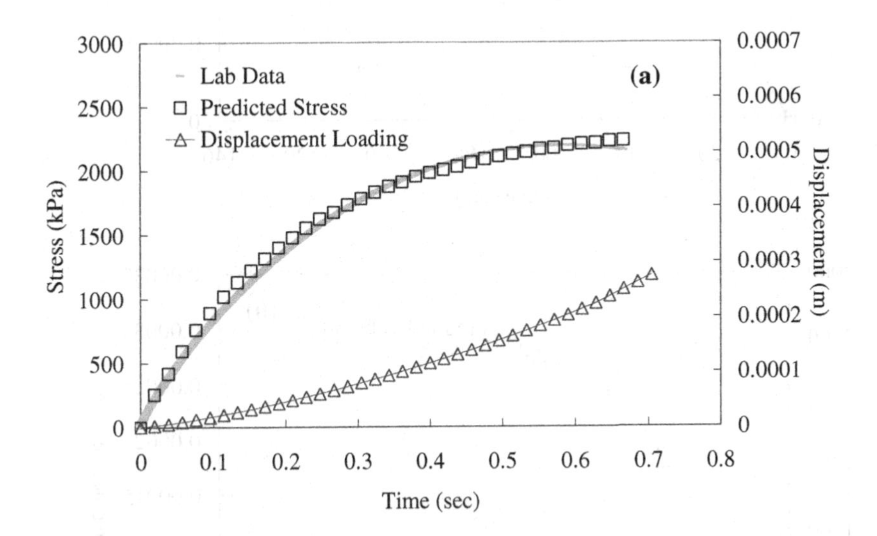

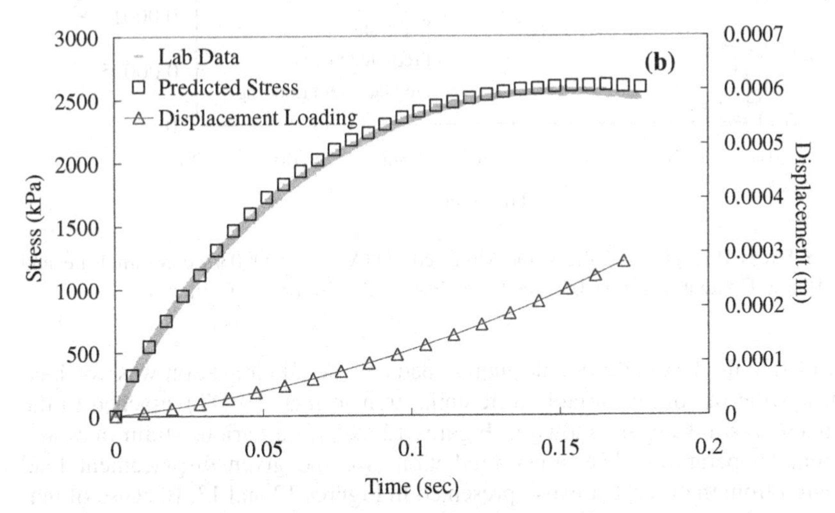

Fig. 12 Stress Prediction of the Unmodified HMA: (a) 0.0063/sec Strain Rate; (b) 0.0294/sec Strain Rate

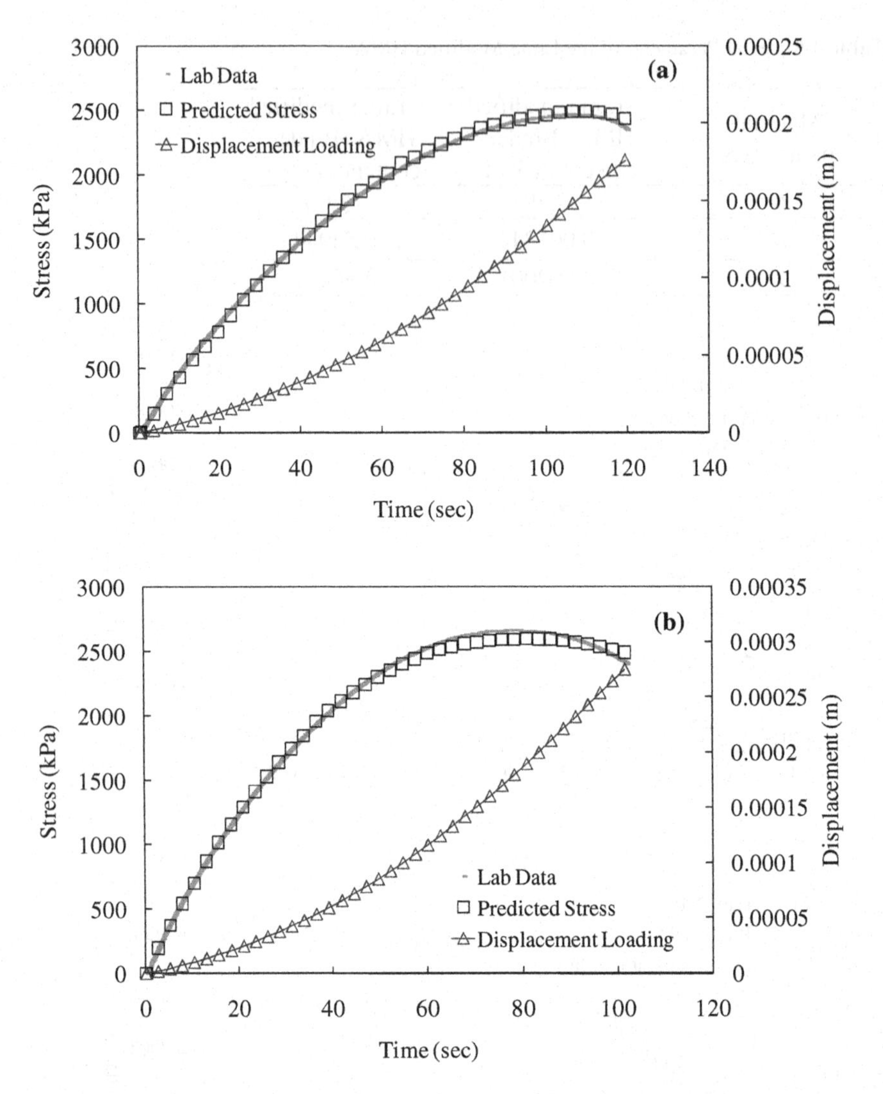

Fig. 13 Stress Prediction of the Lime-Modified HMA: (a) 0.00003/sec Strain Rate and 5.3% Binder Content; (b) 0.000055/sec Strain Rate and 5.8% Binder Content

connected to the loading frame through a load cell. Axial elongation was obtained by measuring the linear variable differential transformer (LVDT) attached to the specimen. Test results are shown in Figures 12 and 13 at various strain rates and different temperatures. The stress predictions for the given displacement load measured from the LVDT are also presented in Figures 12 and 13. Because of machine compliance (e.g., deformation of certain machine components along the loading train under load), the displacements were measured from the LVDT attached to the tested specimens. As shown in these figures, the stress predictions are accurately matched using the experimental data. Furthermore, the stress response for a given displacement load is sensitive to the strain rate. For example, a faster strain rate results in a larger stress level. The purely viscoelastic responses of the unmodified and lime-modified HMA concretes are shown in Figures 14 and 15. It can be observed that the LVE responses depart from the experimental data at the early stage.

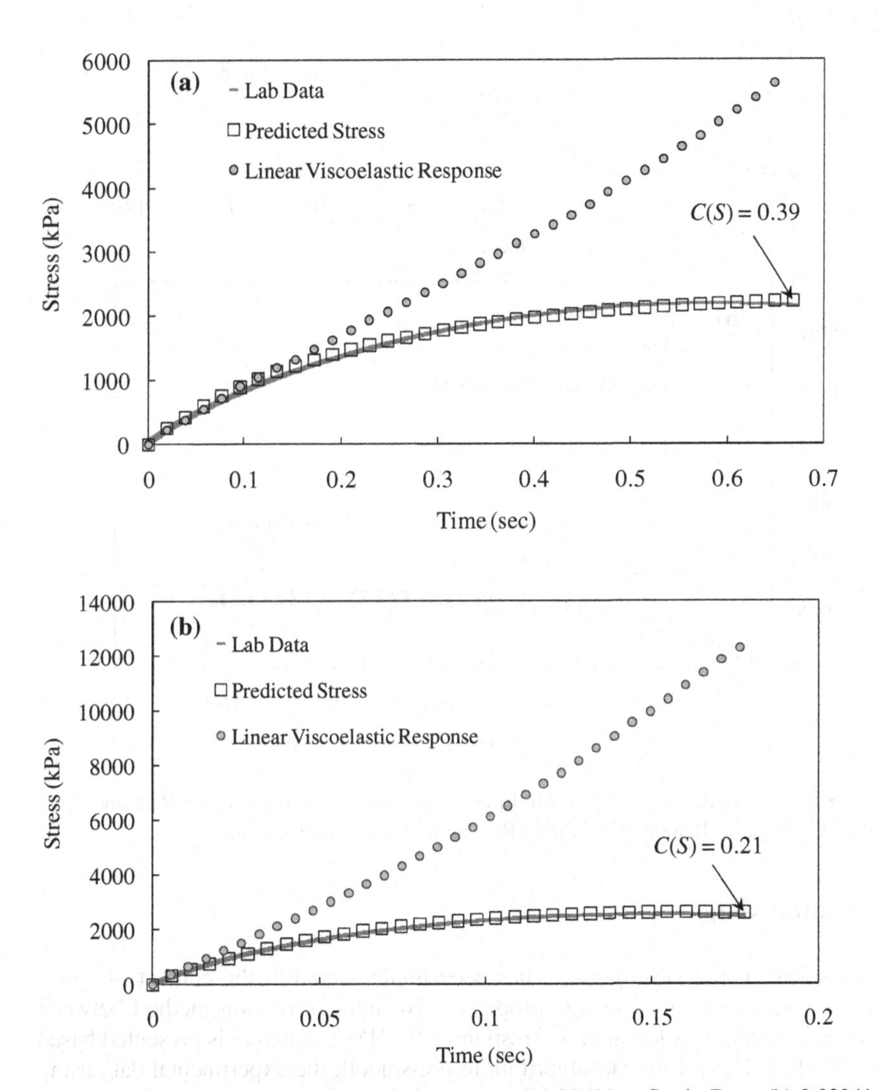

Fig. 14 LVE Response of the Unmodified HMA: (a) 0.0063/sec Strain Rate; (b) 0.0294/sec Strain Rate

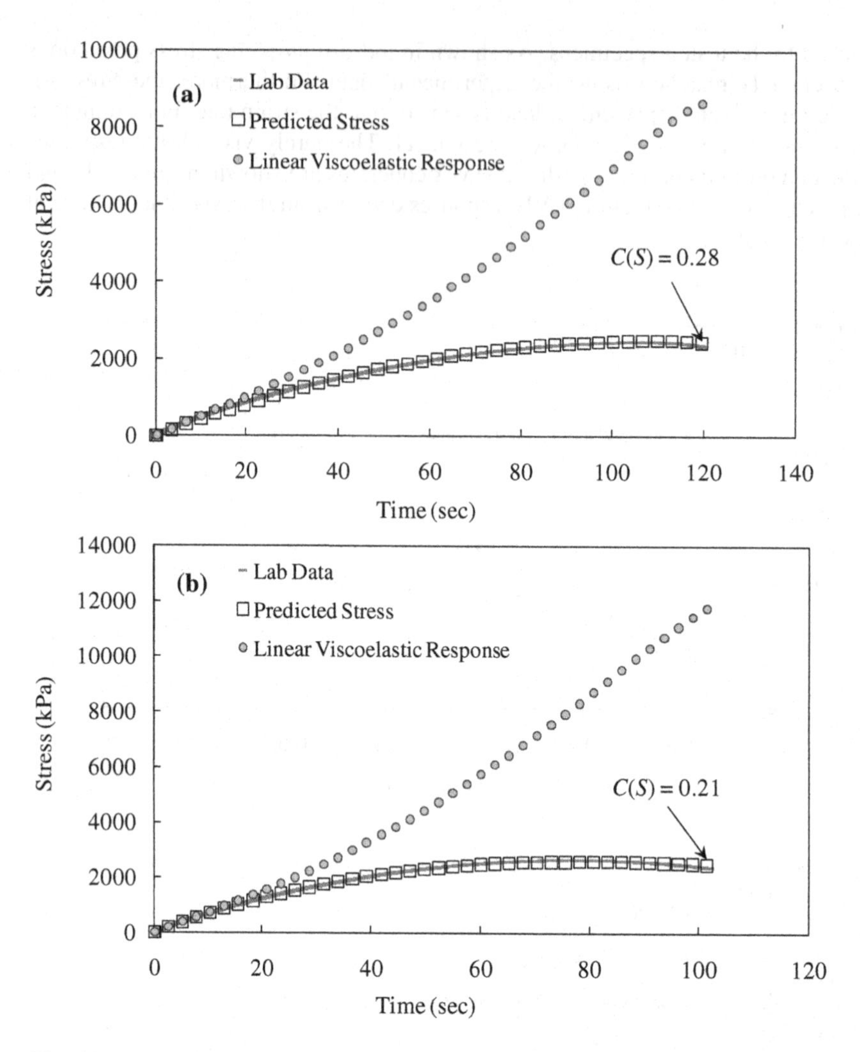

Fig. 15 LVE response of the Lime-Modified HMA: (a) 0.00003/sec Strain Rate and 5.3% Binder Content; (b) 0.000055/sec Strain Rate and 5.8% Binder Content

5 Conclusions

In this chapter, the HS algorithm has been implemented in the context of determining viscoelastic and damage properties. An interconversion method between time- and frequency- domain LVE responses for HMA concrete is presented based on a methodology of the HS algorithm to pre-smooth the experimental data using the log-sigmoidal function before fitting to a Prony series. Also, in order to model the material response of HMA concrete to different strain rate loadings, which induce damage growth, the damage function can be characterized by fitting experimental results using the HS algorithm. Through the application to the prediction of the damage behavior of HMA concrete, the stress predictions result in accurate matches using the experimental data.

References

1. Tschoegl, N.W.: Time dependence in material properties: an overview. Mechanics of Time-Dependent Materials 1, 3–31 (1997)
2. Lee, S., Knauss, W.G.: A note on the determination of relaxation and creep data from ramp tests. Mechanics of Time-Dependent Materials 4, 1–7 (2000)
3. Kaloush, K.: Simple Performance Test for Permanent Deformation of Asphalt Mixtures, Ph.D. Dissertation, Arizona State University, Tempe, AZ (2001)
4. Kim, Y.R., Lee, H.J., Little, D.N.: Fatigue characterization of asphalt concrete using viscoelasticity and continuum damage theory. Journal of the Association of Asphalt Paving Technologists 66, 520–569 (1997)
5. Daniel, J.S., Kim, Y.R.: Development of a simplified fatigue test and analysis procedure using a viscoelastic continuum damage model. Journal of Association of Asphalt Paving Technologists 71, 619–650 (2002)
6. Chehab, G.R., Kim, Y.R., Schapery, R.A., Witczak, M.W., Bonaquist, R.: Time-temperature superposition principle for asphalt concrete mixtures with growing damage in tension. Journal of Association of Asphalt Paving Technologists 71, 559–593 (2002)
7. Geem, Z.W., Kim, J.H., Loganathan, G.V.: A new heuristic optimization algorithm: harmony search. Simulation 76(2), 60–68 (2001)
8. Lee, K.S., Geem, Z.W.: A new meta-heuristic algorithm for continues engineering optimization: harmony search theory and practice. Computer Methods in Applied Mechanics and Engineering 194, 3902–3933 (2004)
9. Goldberg, D.E.: Genetic Algorithms in Search Optimization and Machine Learning. Addison-Wesley, Reading (1989)
10. Lee, H.J., Kim, Y.R.: A uniaxial viscoelastic constitutive model for asphalt concrete under cyclic loading. Journal of Engineering Mechanics 124(11), 1224–1232 (1998)
11. Schapery, R.A.: Correspondence principles and a generalized J integral for large deformation and fracture analysis of viscoelastic media. International Journal of Fracture 25, 195–223 (1984)
12. Schapery, R.A.: Theory of mechanical behavior of elastic media with growing damage and other changes in structure. Journal of the Mechanics and Physics of Solids 38, 215–253 (1990)
13. Park, S.W., Kim, Y.R., Schapery, R.A.: A viscoelastic continuum damage model and its application to uniaxial behavior of asphalt concrete. Mechanics of Material 24, 241–255 (1996)
14. Asphalt Institute, SP-2 Superpave Mix Design, Asphalt Institute (2001)
15. Schapery, R.A., Park, S.W.: Methods of interconversion between linear viscoelastic material functions. Part II: an approximate analytical method. International Journal of Solids and Structures 36(11), 1677–1699 (1999)
16. Park, S.W., Kim, Y.R.: Fitting Prony-series viscoelastic models with power-law presmooting. Journal of Materials in Civil Engineering 13, 26–32 (2001)
17. Biot, M.A.: Theory of stress-strain relations in anisotropic viscoelasticity and relaxation phenomena. Journal of Applied Physics 25, 1385–1391 (1954)

18. Widder, D.V.: An Introduction to Transformation Theory. Academic, New York (1971)
19. Bažant, Z.P., Xi, Y.: Continuous retardation spectrum for solidification theory of concrete creep. Journal of Engineering Mechanics 121, 281–288 (1995)
20. Zi, G., Bažant, Z.P.: Continuous relaxation spectrum for concrete creep and its incorporation into microplane model M4. Journal of Engineering Mechanics 128, 1331–1336 (2002)
21. Kim, Y.R., Lee, Y.C., Lee, H.J.: Correspondence principle for characterization of asphalt concrete. Journal of Materials in Civil Engineering 7(1), 59–68 (1995)
22. Christensen, R.M.: Theory of Viscoelasticity: An Introduction 2nd. Academic Press, New York (1982)
23. Coleman, T.F., Li, Y.: An interior trust region approach for nonlinear minimization subject to bounds. SIAM J. Optimization 6, 418–445 (1996)

Author Index